Conrad Wilson

2002

This volume is one of a series of short biographies derived from *The New Grove Dictionary of Music and Musicians, second edition* (London, 2001). The four volumes that inaugurate this series were chosen by John Tyrrell as outstanding examples of the biographical articles in the new edition; they are printed here with little alteration.

Laura Macy
London, 2001

THE NEW GROVE®

HAYDN

James Webster and Georg Feder

GROVE

MACMILLAN PUBLISHERS LIMITED, LONDON

PALGRAVE, NEW YORK, NY

First published in
The New Grove Dictionary of Music and Musicians®, second edition
edited by Stanley Sadie, 2001

The New Grove and *The New Grove Dictionary of Music and Musicians*
are registered trademarks of Macmillan Publishers Limited, London,
and its associated companies

First published in the UK 2002 by Macmillan Publishers Limited, London

This edition is distributed within the UK and Europe
by Macmillan Publishers Limited, London.

First published in North America in 2002 by Palgrave,
175 Fifth Avenue, New York, NY

Palgrave is the new global publishing imprint of St. Martin's Press LLC Scholarly and
Reference Division and Palgrave Publishers Ltd. (formerly Macmillan Press Ltd.)

British Library Cataloguing in Publication Data
The New Grove Haydn (The New Grove composer biographies series)
  1. Haydn, Joseph, 1732–1809 2. Composers – Austria – Biography
  I. Sadie, Stanley, 1930– II. Tyrrell, John
  780.9'2

  ISBN 0-333-80407-4

Library of Congress Cataloguing-in Publication Data
The New Grove Haydn : the New Grove composer biographies / edited by
  James Webster and Georg Feder
    p. cm. - (Grove music)
  Includes biographical references and index.
  ISBN 0-312-23323-X (pbk.)
  1. Haydn, Joseph, 1732–1809. 2. Composers–Austria–Biography.
  I. Title: New Grove composer biographies. II. Webster, James, 1942–
  III. Feder, Georg. IV. Series.
  ML410.H4 N48 2000
  780'.92–dc21
  [B]   00-031123

Contents

Abbreviations

General

A – alto
Acc. – accompaniment, accompanied by
addl – additional
ad lib – ad libitum
AMS – American Musicological Society
anon. – anonymous
appx(s) – appendix(es)
attrib. – attribution(s), attributed to; ascription(s), ascribed to
Aug – August
aut. – autunm
b – born
b – bass
B – bass [voice]
bap. – bapitzed
bc – basso continuo
bn – bassoon
Bs – Benedictus
c – circa
cant(s). – cantata(s)
cf – confer [compare]
chap(s). – chapter(s)
cl – clarinet
conc(s). – concerto(s)
CT – Connecticut
d – died
DC – District of Columbia
db – double bass
dbn – double bassoon
Dec – December
ded(s). – dedication(s), dedicated to
dg – *dramma giocoso*
diss. – dissertation
edn(s) – edition(s)
eng – English
facs. – facsimile
Feb – February
fig(s). – figure(s) [illustration(s)]
fl – flute
frag(s). – fragment(s)
Ger. – German
Gl – Gloria

grad(s) – gradual(s)
gui – guitar
hn – horn
hp – harp
hpd – harpsichord
ibid. – ibidem [in the same place]
IN – Indiana
inc. – incomplete
incid – incidental
incl. – includes, including
insts – instruments
int(s) – intermezzo(s), introit(s)
It. – Italian
Jan – January
Jb – Jahrbuch [yearbook]
Jg – Jahrgang [year of publication/ volume]
jr – junior
kbd – keyboard
Ky – Kyrie
lib(s) – libretto(s)
MA – Massachusetts
MS(S) – manuscript(s)
movt(s) – movement(s)
NC – North Carolina
n.d. – no date of publication
NJ – New Jersey
no(s). – number(s)
Nov – November
ob – *opera buffa*, oboe
obbl – obbligato
oc – *opéra comique* [genre]
Oct – October
off(s) – offertory (offertories)
op(s) – opera(s)
op(p). – opus, opera [plural of opus]
orat(s) – oratorio(s)
orch – orchestra(tion), orchestral
org – organ
orig. – original(ly)
os – *opera seria*
ov(s). – overture(s)

p(p). – page(s)
perc – percussion
perf(s). – performance(s), performed (by)
pic – piccolo
Ps(s) – Psalm(s)
pt(s) – part(s)
pubd – published
pubn(s) – publication(s)
qt(s) – quartet(s)
qnt – quintet
R – photographic reprint [edition of score or early printed source]
recit(s) – recitative(s)
rept. – reprinted
rev(s). – revision(s), revised (by/for)
RISM – Répertoire International des Sources Musicales
S – soprano
Sept – September
ser. – series
Spl – Singspiel
str – string(s)
summ. – summer
suppl(s). – supplement(s), supplementary
sym(s). – symphony, symphonies
T – tenor
timp – timpani
tpt – trumpet
trans. – translation
trbn – trombone
transcr(s). – transcription(s), transcribed by/for
U. – University
unpubd – unpublished
v – voice
va – viola
vc – cello
vle – violone
vn – violin
vv – voices
WI – Wisconsin

Bibliographic

19CM – *19th Century Music*
AcM – *Acta musicologica*
AMw – *Achiv für Musikwissenschaft*
AMZ – *Allgemeine musikalische Zeitung* (1798–1848, 1863–5, 1866–82)
AnM – *Anuario musical*
AnMc, AnMc – *Analecta musicologica*
BeJb – *Beethoven-Jahrbuch*
BMw – *Beiträge zur Musikwissenschaft*
CMc – *Current Musicology*
EMc – *Early Music*
FAM – *Fontes artis musicae*
GerberNL – E.L. Gerber: *Neues historisch-biographisches Lexikon der Tonkünstler*
GfMKB – *Gesellschaft für Musikforschung: Kongress-Bericht* [1950–]
IMSCR – *International Musicological Society: Congress Report* [1930–]

ImusSCR – *International Musical Society: Congress Report* [II–IV, 1906–11]
JAMS – *Journal of the American Musicological Society*
JbMP – *Jahrbuch der Musikbibliothek Peters*
JM – *Journal of Musicology*
JMR – *Journal of Musicological Research*
JMT – *Journal of Music Theory*
JRMA – *Journal of the Royal Musical Association*
KJb – *Kirchenmusikalisches Jahrbuch*
Mf – *Die Musikforschung*
MGG1 – *Die Musik in Geschichte und Gegenwart*
MJb – *Mozart-Jahrbuch* [Salzburg, 1950–]
ML – *Music & Letters*
MQ – *Musical Quarterly*
MR – *Music Review*

MT – *Musical Times*
NbeJb – *Neues Beethoven-Jahrbuch*
NewmanSCE – W.S. Newman: *The Sonata in the Classical Era* (Chapel Hill, NC, 1963, 3/1983)
NOHM – *The New Oxford History of Music* (Oxford, 1954–90)
ÖMz – *Österreichische Musikzeitung*
PRMA – *Proceedings of the Royal Musical Association*
RBM – *Revue belge de musicologie*
RdM – *Revue de musicologie*
SIMG – *Sammelbände der Internationalen Musik-Gesellschaft*
SMH – *Sudia musicologica Academiae scientiarum hungaricae*
SMw – *Studia zur Musikwissenschaft*
ZfM – *Zeitschrift für Musik*
ZMw – *Zeitschrift für Musikwissenschaft*
ZT – *Zenetudományi tanulmányok*

Library sigla

A-El – Austria, Eisenstadt, Burgländisches Landesmuseum
A-HE – Austria, Heiligenkreuz, Zisterzienserkloster
A-SEI – Austria, Seitenstretten, Benediktinerstift, Musikarchiv
A-Sm – Austria, Salzburg, Internationale Stiftung Mozarteum, Bibliotheca Mozartiana
A-ST – Austria, Stams, Zisterzienserstift, Musikarchiv
A-Wgm – Austria, Vienna, Gesellschaft der Musikfreunde
A-Wn – Austria, Vienna, Österreichische Nationalbibliothek, Musiksammlung
A-Wst – Austria, Vienna, Stadt- und Landesbibliothek, Musiksammlung

CZ-Bm – Czech Republic, Brno, Moravské Zemské Muzeum, Oddělení Dějin, Hudby
CZ-KRa – Czech Republic, Kroměříž, Státní y Zámek a Zahrady, Historicko-Umělecké Fondy, Hudební Archív
CZ-Pnm – Czech Republic, Prague, Národní Muzeum
D-Bsb – Germany, Berlin, Staatsbibliothek zu Berlin Preussischer Kulturbesitz
D-HR – Germany, Harburg (nr Donauwörth), Fürstlich Oettingen-Wallerstein'sche Bibliothek Schloss Harburg
D-Lem – Germany, Leipzig, Leipziger Städtische Bibliotheken, Musikbibliothek

D-Tl – Germany, Tübingen, Schwäbisches Landesmusikarchiv
H-Bn – Hungary, Budapest, Országos Széchényi Könyvtár
H-Gk – Hungary, Győr, Káptalan Magánlevéltár Kottatára
J-Tn – Japan, Tokyo, Nanki Ongaku Bunko
PL-Kj – Kraków, Univwersytet Jagielloński, Biblioteka Jagiellońska

List of Illustrations

Joseph Haydn began his career in the traditional patronage system of the late Austrian Baroque, and ended as a 'free' artist within the burgeoning Romanticism of the early 19th century. Famous as early as the mid-1760s, by the 1780s he had become the most celebrated composer of is time, and from the 1790s until his death was a culture-hero throughout Europe. Since the early 19th century he has been venerated as the first of the three 'Viennese Classics' (Haydn, Mozart, Beethoven). He excelled in every musical genre; during the first half of his career his vocal works were as famous as his instrumental ones, although after his death the reception of his music focussed on the latter (except for *The Creation*). He is familiarly known as the 'father of the symphony' and could with greater justice be thus regarded for the string quartet; no other composer approaches his combination of productivity, quality and historical importance in these genres. In the 20th century he was understood primarily as an 'absolute' musician (exhibiting wit, originality of form, motivic saturation and a 'modernist' tendency to problematize music rather than merely to compose it), but earnestness, depth of feeling and referential tendencies are equally important to his art.

---

1. CHILDHOOD AND YOUTH 1732–c1749

Documentary information on Haydn's life and musical activity before his employ by the Esterházy court in 1761 is scanty. The primary sources comprise an autobiographical letter of 1776 and brief biographies published just after his death by (in order of general reliability) Georg August Griesinger, Albert Christoph Dies, Giuseppe Carpani and Nicolas Etienne Framery, supplemented by parish registers, musical archives, dated autographs and the like. Franz Joseph Haydn (neither he nor his contemporaries used the name Franz) was born on 31 March 1732 in Rohrau, Lower Austria, into a family of primarily South German stock, albeit in an area of considerable ethnic diversity in which Croats and Hungarians were also prominent. His immediate ancestors were not peasants (as legend has it), but artisans and tradespeople. His grandfather and his father, Mathias (1699–1763), were master wheelwrights; Mathias also functioned as *Marktrichter* (magistrate) of the 'market village' (as Haydn called it)

1

Rohrau, near Bruck an der Leitha. Rohrau was a possession of Count Karl Anton Harrach (1692–1758); his grandson Karl Leonhard (1765–1831) erected a monument to Haydn in the castle garden in 1793. Haydn's mother, Anna Maria Koller (1707–54), had before her marriage in 1728 been a cook at the Harrach castle.

Mathias Haydn was 'a great lover of music by nature' (this phrase in Haydn's laconic account has occasionally been taken as applying to Harrach, but it is his father who was meant), who 'played the harp without reading a note of music'; his mother sang the melodies. Indeed all three of their surviving male children became professional musicians, two of them famous composers. (Joseph was the second of 12 children, Michael, 1737–1806, the sixth. The eleventh, Johann Evangelist, 1743–1805, was a tenor in a church choir and later at the Esterházy court.) Dies says of Haydn's father that 'all the children had to join in his concerts, to learn the songs, and to develop their singing voice', adding that he also organized concerts among the neighbours.

Haydn's talent became evident early on. 'As a boy of five I sang all [my father's] simple easy pieces correctly'; according to Griesinger he still remembered these melodies in old age. 'Almighty God ... granted me so much facility, especially in music, that when I was only six I boldly sang masses down from the choirloft, and could also get around on the harpsichord and violin'. In 1737 or 1738 Johann Mathias Franck, a cousin of Mathias Haydn's by marriage and a school principal in the nearby town of Hainburg (Mathias's birthplace), heard Haydn sing in the family circle; Griesinger and Dies also have him pretending to be playing a violin by scraping a stick against his arm. Franck was so impressed by Haydn's voice and musical accuracy that he suggested that he come to live with him, 'so that there I could learn the rudiments of music along with other juvenile necessities'. It being clear that his abilities could not be developed in Rohrau, his parents agreed, whether in the hope that he might amount to something as a musician or the belief that musical and educational accomplishments might be useful in what they (especially his mother) imagined as his true calling, that of a priest.

Franck was not only a school principal but the choir director of a Hainburg church; presumably he oversaw Haydn's education personally. The latter was scarcely an autodidact, as myth used to have it. Griesinger writes:

> He received instruction in reading and writing, in the catechism, in singing, and on almost all the string and wind instruments, and even on the timpani: 'I will be grateful to this man even in the grave',

Haydn often said, 'that he taught me so much, even though in the process I received more beatings than food'.

Such exaggerations aside, he doubtless made rapid progress; his account of mass singing and harpsichord and violin studies 'in my sixth year' implies that these took place in Hainburg. As Griesinger says, his schooling was not musical alone; this was also the case when he was a choirboy in Vienna, where his non-musical studies, though 'scanty', included Latin, religion, arithmetic and writing.

In 1739 or 1740 ('in my 7th year'; Griesinger and Dies: in his eighth year) Haydn was recruited to serve as choirboy at the Stephansdom in Vienna: 'Kapellmeister Reutter, on a trip through Hainburg, heard my thin but pleasant voice from a distance, and at once accepted me into the *Capell Hauss*' (choir school). Georg Reutter the younger, Kapellmeister at the Stephansdom since 1738 (later Hofkapellmeister), was travelling through the provinces in search of new talent; in Hainburg the parish priest, an old friend, suggested that Haydn might be a suitable candidate. According to several accounts Haydn did not know how to trill but, after Reutter demonstrated, triumphantly got it right on his third attempt, thus sealing his acceptance. For the next ten years, 'I sang soprano both at St Stephan's and at court to great applause'. At the choir school, 'I was taught the art of singing, the harpsichord and the violin by very good masters'; in singing these included Adam Gegenbauer and the tenor Ignaz Finsterbusch (both *d* 1753). To be sure, there was apparently little formal training in theory or composition, although the singing included *solfeggio* and the harpsichord instruction probably entailed figured bass. But in their enthusiasm for the notion that Haydn's development amounted to 'making something out of nothing' (Dies, allegedly quoting Haydn), most accounts again exaggerate this supposed lack of instruction. 'Haydn recalled having had only two lessons [in theory] from the worthy Reutter', writes Griesinger, but if he could recall two, he might have had more. In any case, 'Reutter encouraged him to make whatever variations he liked on the motets and Salves that he had to sing in church, and this discipline soon led him to ideas of his own, which Reutter corrected'; this scarcely implies outright neglect.

It was surely not on Haydn's own that 'he also came to know Mattheson's *Der vollkommene Capellmeister* (1739) and Fux's *Gradus ad Parnassum* (1725). With tireless exertion Haydn sought to understand Fux's theory; he worked his way through the entire treatise'. However, although both Griesinger and Dies mention Fux in the context of the choir school, Haydn's study of him would more plausibly have taken place during the 1750s. In any case, his copy of *Gradus* is heavily annotated in Latin; he made it the basis of his own teaching of composition, as did Mozart. Another activity entrusted to competent older choirboys was the instruction of their younger colleagues in musical fundamentals; among those whom Haydn taught was his brother Michael, who joined him there about 1745. Most important, for ten full years, at a highly impressionable age, Haydn rehearsed and sang in performances of the greatest art-music then being produced in Catholic Europe, amid the pomp and splendour of the cathedral and court of an imperial capital. This experience will have fundamentally shaped his musical intellect even without formal training in composition.

But this life could not last; his voice broke. A characteristic anecdote adds insult to injury by relating that after one performance Maria Theresa said that he sang 'like a crow', while rewarding Michael for his beautiful singing. Griesinger states that Reutter had earlier suggested that Haydn might become a castrato, but his father refused permission (although this seems potentially inconsistent with his parents' original hope that he become a priest). Be this as it may, soon after his voice broke he was dismissed from the choir school. Haydn wrote that he remained there 'until into my 18th year' (i.e. April 1749 to March 1750); Griesinger's estimate, 'in his sixteenth year', is generally thought to be too early. Carl Ferdinand Pohl, who had access to many documents now lost but gives no source in this instance, writes: 'We find Haydn on the street; it was a damp November evening in 1749'. Pathos aside (year and atmosphere derive from Framery), the date is consistent with Haydn's statement.

2. VIENNA, *c*1750–61

Haydn's account of his freelance 1750s narrates a classic 'rags to riches' story:

> When my voice finally broke, for eight whole years I was forced to eke out a wretched existence by teaching young people. Many geniuses are ruined by this miserable [need to earn their] daily bread, because they lack time to study. This could well have happened to me; I would never have achieved what little I have done, had I not carried on with my zeal for composition during the night. I composed diligently, but not quite correctly, until I finally had the good fortune to learn the true fundamentals of composition from the famous Porpora (who was in Vienna at the time). Finally, owing to a recommendation from the late [Baron] von Fürnberg

(who was especially generous to me), I was appointed as director
with Count Morzin, and from there as Kapellmeister with his high-
ness Prince [Esterházy].

This period comprises three stages, of which the first two overlap without
clear division. During the 'lean years', about 1749 to the mid-1750s,
Haydn was a freelance musician, teacher and budding composer. Even then,
however, he was reaping professional and social advantage from contact
with figures such as Porpora and Metastasio. Beginning around 1753,
and increasingly after 1755, his compositional activity expanded, as his reputation and access to patronage grew. His first regular appointment, as
director of music for Count Morzin, began probably in 1757 and lasted
until the winter of 1760–61 or the spring of 1761.

Haydn's first lodgings (according to Framery) were offered by Johann
Michael Spangler, a tenor (later *regens chori*) at the Michaelerkirche, in a
garret with his wife and infant son (*b* February 1749). This solution
obviously could be no more than temporary, especially as Spangler's wife
was soon pregnant with their second child, Maria Magdalena (*b* 4
September 1750). (In 1768 Haydn engaged Maria Magdalena Spangler as a
soprano at the Esterházy court, where among other roles she created
Vespina in *L'infedeltà delusa* and Rezia in *L'incontro improvviso*; she was also
the first Sara in *Il ritorno di Tobia*.) Another good deed was done
him by Johann Wilhelm Buchholz, a lacemaker, whose granddaughter was
remembered in Haydn's will 'because her grandfather lent me 150 gulden
without interest in my youth and great need'; the amount was close to a
year's salary for an ordinary musician at a minor court. It was perhaps in
the following spring (1750) that he journeyed to the huge Benedictine
pilgrimage church in Mariazell (Styria). Griesinger relates that he took with
him 'several motets which he had composed and asked the *regens chori* there
for permission to put out the parts in the church and sing them', and continues with an anecdote according to which Haydn the next day got his
way by trickery. If 'motet' means a liturgical work other than a mass, it can
only have been his first *Lauda Sion* hymns, HXXIIIc:5; another possibility
is the *Missa brevis* in F. In any case this pilgrimage was important to
Haydn; later he composed both the *Missa Cellensis* (HXXII:5) and the
'Mariazellermesse' (HXXII:8)with Mariazell in mind.

According to Pohl, it was in the spring or summer of 1750 that
Haydn occupied his most frequently described early lodgings: a 'miserable
little garret without a stove' (Griesinger) in the so-called Michaelerhaus,
attached to the Michaelerkirche. At this time 'his entire life was devoted to
giving lessons, the study of his art, and performing. He played in serenades

and in orchestras for pay, and devoted himself diligently to composition, for "when I sat at my old, worm-eaten clavier, I envied no king his good fortune"'. Here occurred the first of many strokes of luck through which, in addition to his genius and unremitting labour, he gradually improved his professional lot. Griesinger writes:

> In the same house ... lived as well the famous poet Metastasio. He was raising one Fräulein Martinez; Haydn was engaged to give her lessons in singing and on the clavier, in return for which he received free board for three years. At Metastasio's he also made the acquaintance of the ageing Kapellmeister Porpora. Porpora was teaching singing to the mistress of the Venetian ambassador, Correr; however, because he was too proper and too fond of his ease to accompany at the piano himself, he delegated this task to our Giuseppe. 'There was no lack of *Ass, Blockhead, Rascal* and pokes in the ribs, but I willingly put up with it all, for I profited immensely from Porpora in singing, composition and Italian.' In the summer Correr travelled with the lady to the popular bathing resort Mannersdorf ... ; Porpora went as well ... and took Haydn with him. For three months Haydn served there as Porpora's valet; he ate at Correr's officers' table, and was paid six ducats [approx. 25 gulden] a month. From time to time he was required to accompany Porpora on the clavier at one Prince von Hildburghausen's, in the presence of Gluck, Wagenseil and other famous masters; the approval of these connoisseurs was especially encouraging to him.

Access to such personages – whose overlapping relations were as much social as artistic – was essential for an aspiring young musician. 'Fräulein Martinez' was the composer and singer Marianne von Martínez. At the court of Joseph Friedrich, Prince of Sachsen-Hildburghausen (1702–87), Haydn could also have encountered Dittersdorf (whom he certainly knew by the mid-1750s) and Giuseppe Bonno, later Hofkapellmeister.

All these events took place during the first half of the 1750s. Haydn's instruction of Martínez began in either 1751 or 1752; presumably his three years with Metastasio were between 1751 and 1754. Porpora arrived in Vienna from Dresden in 1752 or early 1753; Haydn might well have met him in March 1753, when Metastasio was considering him as composer of his new opera *L'isola disabitata* (which in the event he assigned to Bonno; Haydn himself set this libretto in 1779). Given the mastery shown in Haydn's music by about 1756, 1753 or 1754 are the latest plausible dates for his having 'learnt the true fundamentals of composition' from Porpora, whose expert knowledge of singing and Italian – 'singing' in this context implies Italian opera and oratorio – was also of great importance; Haydn became fluent in Italian and the italianate singing style. In addition, it may well have been at Porpora's instigation that he systematically worked through Fux's *Gradus ad Parnassum* (the only work mentioned by any source that offers 'true fundamentals'). Another important musical encounter was Haydn's discovery of Carl Philipp Emanuel Bach, but this is unlikely to have taken place as early as about 1750, as the biographers claim. Dies portrays Haydn asking for 'a good theoretical textbook'; this can refer only to the second volume of Bach's *Versuch über die wahre Art das Clavier zu spielen*. However, it appeared far too late (1762) to serve the function Dies attributes to it; even Bach's first volume (1753) was apparently not sold in Vienna until the 1760s. Moreover, unlike Fux or Mattheson, neither volume figures in Haydn's library catalogue (1804) or his estate. Indeed Griesinger speaks more plausibly of compositions:

> About this time [his move to the Michaelerhaus] Haydn came upon the first six sonatas of Emanuel Bach; 'I did not leave my clavier until I had played them through, and whoever knows me thoroughly must discover that I owe a great deal to Emanuel Bach, that I understood him and studied him with diligence. Emanuel Bach once paid me a compliment on that score himself'.

Although it is unclear which of Bach's sonatas Griesinger meant by 'the first six', there is no doubt of his influence on Haydn as a composer. Again, however, Haydn's style does not reflect that influence until the 1760s.

An important early personal contact was with Joseph Felix von Kurz, a well-known comic actor (under the stage-name Bernardon) and minor impresario active at the Kärntnertortheater, for whom Haydn supplied music to *Der (neue) krumme Teufel*, a comedy of the 'Hanswurst' type. It was apparently given its première in the 1751–2 season and revived in May 1753, with considerable success. Neither libretto nor music of this, his earliest stage work, survives; a libretto does survive for a later version of c1759, often called *Der neue krumme Teufel*, but, again, there is no music. It has been speculated that many anonymous numbers in contemporary Viennese collections of theatrical songs stem from this work or others that Haydn might have composed, although document is lacking.

Haydn's lot improved substantially in the mid-1750s, as Griesinger describes:

> At first Haydn received only two gulden a month for giving lessons, but gradually the price rose to five gulden, so that he was able to look about

> for more suitable quarters. While he was living in the Seilerstätte, all his few possessions were stolen ... Haydn soon saw his loss made good by the generosity of good friends ... [he] recovered through a stay of two months with Baron Fürnberg, which cost him nothing.

A 150% increase in fees implies a rise not only in Haydn's economic status but his professional reputation – and therefore increased access to patronage. The most important figure was Baron Carl Joseph Fürnberg (1720–67), who employed him as music master to his children (he lived near the Seilerstätte), commissioned his first string quartets and eventually recommended him to Morzin. Important as well was the elder Countess Maria Christine Thun, who (according to Framery) took singing and keyboard lessons from Haydn.

His freelance activities continued apace. Griesinger writes:

> In this period Haydn was also leader of the orchestra in the convent of the Barmherzige Brüder ... at 60 gulden a year. Here he had to be in the church at eight o'clock in the morning on Sundays and feast days, at ten o'clock he played the organ in what was then the chapel of Count Haugwitz, and at eleven o'clock he sang at St Stephan's. He was paid 17 kreutzer for each service. In the evenings Haydn often went 'gassatim' with his musical comrades, when one of his compositions was usually performed, and he recalled having composed a quintet [possibly HII:2] for that purpose in 1753.

(Both Griesinger and Dies supply the obligatory comic anecdotes involving Haydn and Dittersdorf.) From 1754 to 1756 Haydn performed as a singer in the Hofkapelle during Lent (one gulden per service, not 17 kreutzer), and in the 1755–6 season as an orchestral violinist for balls during Carnival (four gulden per evening). In the Hofkapelle he sang both concerted and *a cappella* works, including Palestrina's *Stabat mater* and the Allegri *Miserere*. He worked for the Barmherziger Brüder (in the second district) *c*1755–8, and for Count Haugwitz probably *c*1756–7 at the 'Theresienkapelle' in the Bohemian chancellery in the Wipplingerstrasse. According to an account by a Prussian prisoner of war, he participated in chamber-music parties arranged by Count Harrach at Rohrau. Of Haydn's many students during these years, Martínez has already been mentioned; another of more than marginal importance was Robert Kimmerling, later *regens chori* at Melk.

Although a sizeable number of Haydn's works originated during the 1750s, documented dates are few. Both the very early *Missa brevis* in F and the first *Lauda Sion* exhibit technical faults, implying that he composed them before learning the 'true fundamentals' (i.e. before *c*1753–4); such faults are found in no other surviving genuine works. Griesinger writes: 'In addition to performing and teaching, Haydn was indefatigable in composing. Many of his easy clavier sonatas, trios and so on belong to this period, and he generally took into consideration the needs and capacities of his pupils'. Numerous tiny keyboard sonatas and 'concertini' indeed survive, although some authorities argue that the smallest sonatas are not necessarily the earliest or least accomplished, and the concertini appear to date from about 1760; possibly some keyboard trios antedate 1755 as well, although none are so short or simple. In any case Haydn's compositional activity increased exponentially in the mid-1750s. The quintet-divertimento HII:2 survives in a later source dated 1754, and many of his ensemble divertimentos probably date from before 1761. Of the ten or so pre-1780 keyboard trios and the 21 or so string trios, the earliest also may date from the mid-1750s, although others are from the early 1760s. Late in life Haydn dated the autographs of the Organ Concerto in C (HXVIII:1) and the *Salve regina* in E (HXXIIIb:1) '1756'.

The precise dates of the final two stages of Haydn's early 'progress' – Fürnberg and Morzin – also remain uncertain. Griesinger writes:

> The following, purely coincidental circumstance led him to try his hand at the composition of quartets. A certain Baron Fürnberg had an estate in Weinzierl, several stages from Vienna; from time to time he invited his parish priest, his estate manager, and Albrechtsberger (a brother of the well-known contrapuntist) in order to have a little music. Fürnberg asked Haydn to compose something that could be played by these four friends of the art. Haydn, who was then 18, accepted the proposal, and so originated his first quartet [incipit of HIII:1], which, immediately upon its appearance, received such uncommon applause as to encourage him to continue in this genre.

Griesinger's statement that Haydn composed his first quartet at 18, although roughly supported by Dies and Carpani, is far too early. All the circumstantial details, as well as the sheer mastery of Haydn's early quartets, suggest rather the Seilerstätte period, i.e. about 1755–7. Whether HIII:1 was actually Haydn's first quartet, or whether he (or Griesinger) named it simply because it occupied first position in Pleyel's famous edition (1801) and therefore in his own thematic catalogue, cannot be determined. In any case Haydn had not yet adopted the 'opus' format; there are, simply, ten early quartets, of which HIII:1–4, 6 (op.1 nos.1–4, 6) and HII:6 ('op.0') are probably the earliest, HIII:10, 12 (op.2 nos.4, 6) perhaps in the middle, and HIII:7–8 (op.2 nos.1–2) probably the latest, perhaps even 1759–60.

Regarding Haydn's employ by Count Karl Joseph Franz Morzin (1717–83), Griesinger states:

> In the year 1759 Haydn was engaged as music director to Count Morzin in Vienna at a salary of 200 gulden, free lodging and board at the officers' table. Here he was finally able to enjoy the happiness of a care-free existence; he was quite contented. The winters were spent in Vienna, the summers in Bohemia [at Dolní Lukavice, usually referred to as Lukavec] .... As music director in the service of Count Morzin Haydn composed his first symphony [incipit of no.1].

Although Dies agrees regarding the date ('about 27'), Haydn's earliest symphony cannot be as late as 1759: a manuscript source for no.37 is dated 1758 (Carpani's date), implying a date of composition in that year or, more likely, at least a year earlier. Moreover, Haydn himself in old age organized a list of his symphonies according to ten-year periods: 1757–67, 1767–77 etc.; '1757' is so precise that he must have believed that it was the actual year of his first symphony – or, perhaps, of his appointment with Morzin. (The list also appears to confirm Griesinger's identification of the symphony we know as 'no.1' as the earliest.) Finally, if one accepts 1749 as the date of Haydn's dismissal from the Stephansdom and takes literally his 'eight whole years' of 'wretched existence', 1757 is implied as end-point; but the most likely marker of the latter, again, was his appointment with Morzin. Be all this as it may, the free-spending Count Morzin soon had to dissolve his little musical establishment. Although the early biographers again disagree as to the date, Haydn's marriage certificate (26 November 1760) refers to him as 'Music Director with Count v. Morzin', so he probably moved from Morzin to Esterházy without meaningful interruption. Haydn's compositions during the Morzin years include about 15 symphonies; keyboard sonatas (including H XVI:6, probably not later than 1760), trios, divertimentos (including H XIV:11, 1760) and concertos; string trios; partitas for wind-band (including H II:15,1760) and possibly the quartets op.2 nos.1–2.

On 26 November 1760 Haydn was married to Maria Anna Aloysia Apollonia Keller (bap. 9 Feb 1729; d 20 March 1800); the marriage contract, in which he pledged 1000 gulden as a matching sum to her dowry (a common custom), is dated 9 November. The bride was the daughter of the wigmaker Johann Peter Keller, who is said variously to have assisted him in his years of poverty or employed him as a music teacher. The early biographers relate that in the mid-1750s Haydn had fallen in love with her younger sister Therese (b 1733), who however was compelled by her devout parents to enter a nunnery in 1755, taking her formal vows in 1756. It has been speculated that he composed the *Salve regina* in E and perhaps the Organ Concerto in C on the latter occasion. Whether (the speculation continues) out of continuing gratitude to the Kellers or owing to a psychological displacement of his own affections, he married the elder sister four years later. The marriage was an unhappy one (we have only his side of the story) and led to infidelities on both sides; as he said to Griesinger (somewhat illogically): 'My wife was incapable of bearing children, and thus I was less indifferent to the charms of other women'.

3. ESTERHÁZY, 1761–90

With Haydn's move to the Esterházy court, evidence regarding his activities increases a hundredfold. However, its scope is uneven: although the archives are informative regarding theatrical activities, entertainments for noble visitors, personnel, payments for services, petitions etc., they tell us little about day-to-day musical activities, especially in the realm of instrumental music. Many documents and musical sources were destroyed in a fire at Eszterháza castle in 1779, and little correspondence of Haydn's survives until the upswing in his commercial activity beginning in 1780.

VICE-KAPELLMEISTER, 1761–5

The Esterházy family, the richest and most influential among the Hungarian nobility, had long been important patrons of culture and the arts; Prince Paul Anton was a music lover and capable performer. Haydn's predecessor as Kapellmeister, Gregor Joseph Werner, had been appointed in 1728; he was a solid professional who composed church music in the first instance, but also symphonies, trio sonatas and other instrumental works including an entertaining 'Musical Calendar' (1748); in 1804 Haydn honoured his predecessor by publishing six introductions and fugues from his oratorios, arranged for string quartet. Paul, who from 1750 to 1752 was ambassador in Naples and travelled widely elsewhere, collected a large quantity of vocal and instrumental music (he had a catalogue made during the period 1759–62; it lists one symphony by Haydn, acquired in 1760). By about 1760 Werner was becoming infirm and his musical orientation increasingly conservative; Paul set about modernizing and enlarging the

11

establishment, appointing several new musicians before recruiting Haydn and others in 1761.

Haydn's appointment was in the first instance as vice-Kapellmeister; the first and last clauses of his contract address this somewhat delicate situation, while illustrating the Esterházys' concern for the welfare of valued employees:

> ... Whereas
>
> 1mo a Kapellmeister at Eisenstadt named Gregorius Werner has devoted many years of true and faithful service to the princely house, but now, on account of his great age and the resulting infirmities ... is not always capable of performing his duties, therefore said Gregorius Werner, in consideration of his long service, shall continue to serve, as Ober-Kapellmeister. On the other hand the said Joseph Haydn, as vice-Kapellmeister, shall be subordinate to ... said Gregorio Werner, quà Ober-Kapellmeister, in regard to the choral music [*Chor-Musique*] in Eisenstadt; but in all other circumstances where any sort of music is to be made, everything pertaining to the music, in general and in particular, is the responsibility of said vice-Kapellmeister.
>
> 14mo His Highness not only undertakes to retain the said Joseph Haydn in service during this period [three years, renewable], but, should he provide complete satisfaction, he shall also have expectations of the position of Oberkapellmeister.

Although the contract is dated 1 May 1761, Haydn may have begun working for the court earlier that year. Griesinger states that he began on 19 March 1760; this cannot be correct, unless it was an error for 1761 (Dies also names 1760), and the specific date '19' is suspect (the surviving contracts begin on the first of the month). But the Prince was in Vienna in March 1761 (music was performed at the Esterházy palace several times that month); indeed he may have remained there much of the time until his death in March 1762. Moreover, the contracts with several musicians appointed 1 April 1761 include a clause requiring them to obey not only the Kapellmeister but the vice-Kapellmeister, but the latter position did not exist until Haydn's appointment. Hence he may well have selected most or all of the musicians hired from April on; i.e., helped to shape the newly constituted orchestra himself.

Haydn's contract, once thought to be demeaning, is now understood as a standard document of its type; its terms were favourable to a young man of 29 with only one previous position to his credit. He was no servant, but a professional employee or 'house officer'; he received 400 gulden a year, plus various considerations in kind including uniforms and board at the officers' table. He was in charge of the 'Camer-Musique', which comprised not only all instrumental music but secular vocal and stage music as well. He had full authority over the musicians, both professionally and in terms of their behaviour. His duties included responsibility for the musical archives and instruments (including purchase, upkeep and repair), instruction in singing, performing both as leader and as soloist ('because [he] is competent on various instruments') – and, of course, composition:

> 4mo Whenever His Princely Highness commands, the vice-Kapellmeister is obligated to compose such works of music as His Highness may demand; further not to communicate [such] new compositions to anyone, still less allow them to be copied [for others], but to reserve them entirely and exclusively for His Highness; most of all to compose nothing for any other person without prior knowledge and gracious consent.

Despite the immense labour and considerable tribulation this position entailed, Haydn must have known that it was the opportunity of a lifetime. One can well understand the joy and satisfaction conveyed by Griesinger's remark: 'It was still granted to Haydn's father [d 1763] to see his son in the blue and gold uniform of the court, and to hear the prince's many praises of his son's talent'.

Paul Anton, already in uncertain health in 1761, declined rapidly in early 1762 and died on 18 March. Childless, he was succeeded by his brother Nicolaus, an even more enthusiastic musician, who harboured even grander designs for the physical and artistic development of the court. Goethe coined the phrase *das Esterházysche Feenreich* ('the fairy kingdom of the Esterhazys') to describe his display at the coronation of the Holy Roman Emperor in Frankfurt in 1764, which has passed into the literature; in his own day he was called *der Prachtliebende*, 'the Magnificent'. His treatment of Haydn was generous: he raised his salary to 600 gulden, regularly dispensed gold ducats as thanks for the submission of baryton trios and after successful opera productions and, following fires that destroyed Haydn's house in Eisenstadt in 1768 and 1776, paid to have it rebuilt. As a matter of course his musical taste decisively influenced what genres Haydn cultivated at court; whether it affected Haydn's style as well cannot be determined (except in cases like the works for baryton).

The musical ensemble was at first very small, normally comprising 13 to 15 players (of whom many performed on more than one instrument): approximately 6 violins, 1 viola, 1 cello, 1 bass, 2 oboes, 2 horns and 1 bassoon (plus a flute in certain works or movements). Haydn led from the violin; no keyboard continuo was employed except in the theatre. Beginning in the 1770s, the ensemble was gradually enlarged, owing primarily to the increasing importance of the court opera; at its height in the 1780s it counted 22 to 24 members. Especially at first, it was manned largely by virtuosos (including Alois Luigi Tomasini, violin, Joseph Weigl and later Anton Kraft, cello, Carl Franz, horn and baryton), some of whom remained at the court for decades. This situation is reflected in many difficult and exposed passages in Haydn's symphonies, as well as numerous concertos from the 1760s. Indeed symphonies nos.6–8, *Le matin*, *Le midi* and *Le soir* (1761) – among his first compositions in his new position; Dies states that the 'times of day' topic was suggested by the prince – were expressly calculated to show off the new ensemble, both as a whole and in terms of the individual players, all of whom receive solos. But Haydn was also demonstrating his own prowess: although the topics were traditional, the works have no precedent, either generally or in his own output.

During the first half of the 1760s Haydn composed chiefly instrumental music, as far as we know exclusively for performance at court. His most productive genre was the symphony, with about 25 works; in addition to nos.6–8 they include nos.22 ('The Philosopher') and 30 and 31 ('Alleluja', 'Hornsignal'). The concertos include two or three violin concertos, the Cello Concerto in C, a concerto for violone (the first ever composed, as far as is known), two horn concertos and one for two horns, one for flute and perhaps one for bassoon; many of them are lost. Only a few keyboard works are known, primarily trios and quartet divertimentos; there are also a few ensemble divertimentos as well as minuets and other dances. In addition, there were a few large-scale vocal works, usually intended as celebrations of particular occasions: the *festa teatrale Acide* (1762, first performed in January 1763, for the marriage of Anton, the prince's eldest son) and the somewhat mysterious *commedia*, *Marchese (La Marchesa Nespola)* (1762–3; only fragments survive, and three similar works are lost), as well as several cantatas honouring Nicolaus himself, whether on his nameday (*Destatevi o miei fidi*, 1763; *Qual dubbio ormai*, 1764), his return from distant journeys (*Da qual gioia improvvisa*, 1764, from Frankfurt; *Al tuo arrivo felice*, 1767, from Paris; the latter work may be the same as the former) or his convalescence from illness (*Dei clementi*, undated). The only sacred vocal work of consequence is the first of Haydn's two *Te Deum* settings (H XXIIIc:1).

We know little of Haydn's daily routine or that of the musical establishment during these years, or of noteworthy events in his life. His contract required him to appear every morning and afternoon to see if music was desired, although a later document specified that academies were to be given regularly on Tuesdays and Thursdays. From 1762 to 1765 Nicolaus lived primarily in Eisenstadt, with frequent shorter stays at other properties. Haydn and his wife lived in an apartment in the same building as the other musicians, next to the 'Bergkirche', just up the hill from the castle. He was seriously ill in the winter of 1764–5; the following year his brother Johann was engaged, nominally as a tenor but *de facto* charitably supported by Haydn.

KAPELLMEISTER, 1766–90

In late 1765 and early 1766 Haydn's status and activities at the Esterházy court changed radically. First came a series of crises in his relations with the prince. In September 1765 the flautist Franz Sigl accidentally burnt down a house; the chief court administrator, Ludwig Pater Rahier (with whom Haydn often clashed), recommended that Sigl be imprisoned, and Haydn was reprimanded by the prince. Haydn however eloquently defended himself and succeeded in having Sigl's punishment reduced to simple dismissal (indeed he was later rehired). In October, Werner, having just signed his last will, wrote a vituperative letter to the prince in which he accused Haydn of neglecting the instruments and musical archives and the supervision of the singers. In late November or early December Nicolaus again sent Haydn a reprimand (perhaps drafted by Rahier), instructing him to see to these matters and to prepare a catalogue of the archives and instruments of the Chor-Musique. At the end stood the following postscript: 'Kapellmeister Haydn is urgently enjoined to apply himself to composition more zealously than heretofore, and especially to compose more pieces that one can play on the [baryton]'. The baryton was a member of the viol family, on which the performer could 'accompany himself' by plucking a series of sympathetically vibrating strings while also playing normally with the bow; the prince was an accomplished performer. Haydn, though doubtless angry and dismayed, at once began to compose baryton trios in quantity. On 4 January 1766 he submitted three new ones (Nicolaus pronounced himself satisfied and awarded him 12 ducats, while immediately ordering six more), and completed a 'book' of 24 (they were elegantly bound insets) that autumn; two additional books followed by July 1768. Thereafter production dropped off somewhat, though remaining steady into the mid-1770s, for a total of 126 trios plus sundry other works.

A different kind of response (so it is assumed) was Haydn's decision to begin a thematic catalogue of his own compositions, and thereby to refute the prince's charge of non-productivity. This document, misleadingly called the 'Entwurf-Katalog', is of capital importance for our knowledge of Haydn's output from the pre-Esterházy days up to the late 1770s, as well as its chronology. It was laid out in about 1765–6 by, Joseph Elssler, the most important music copyist at the court, doubtless according to Haydn's plan; Haydn made additional entries more or less regularly until the late 1770s.

On 3 March 1766 Werner died; Haydn was now Kapellmeister, responsible for the church music as well as everything else. It was presumably this higher status (his salary did not change) that induced him in May to purchase a house in Eisenstadt (now a Haydn museum). A more important change was signalled later that year: a portion of the court, including Haydn and some musicians, spent the summer at Nicolaus's splendid new castle, Eszterháza, then beginning to rise in reclaimed swampland east of Lake Neusiedl (present-day Hungary). Over time, the prince became increasingly attached to it, and 'summer' eventually expanded to ten months. Such an extension occasioned the 'Farewell' Symphony (no. 45, autumn 1772), in which the pantomime of the departing musicians brought home to the prince the need to return to Eisenstadt.

As a result of these new circumstances, Haydn's compositional activity changed substantially. In addition to the upsurge in baryton music, in 1766 he began to compose large-scale vocal works, both sacred and secular. In the former domain he at once produced two works on the largest scale, with an astonishing assurance of style and technique for someone who had composed no church music for a decade. The first was the *Missa Cellensis in honorem BVM* of 1766 (possibly completed later), apparently intended for Mariazell (where earlier Esterházys had erected a chapel) or a Viennese church associated with that shrine. More masses followed: in 1768 the *Missa Sunt bona mixta malis*, about 1768–9 the *Missa in honorem BVM* (Grosse Orgelmesse; Haydn presumably performed the obbligato organ part), in 1772 the *Missa Sancti Nicolai* (the title implies a celebration of the prince's nameday, 6 December) and about 1775–8 a *Missa brevis* ('Kleine Orgelmesse'). His other 'inaugural' liturgical masterpiece was the *Stabat mater* of 1767. Its original purpose is unknown; Haydn was confident enough to send it to Hasse, earning a letter of praise that he much valued, and he performed it in Vienna in March 1768. There followed a *Salve regina* (H XXIIIb:2; 1771) for four solo voices, string orchestra and obbligato organ (again performed by Haydn). It was presumably this work, not (as he later claimed) the *Stabat mater*, that resulted from his vow to compose a work of thanksgiving to the Virgin if he were cured of a serious illness; he suffered from a 'raging fever' (Griesinger) about 1770–71, so threatening that his brother applied for leave from Salzburg to visit him. The celebratory cantata *Applausus* (1768) was commissioned in honour of the abbot of the Cistercian monastery at Zwettl; because Haydn was unable to be present, he accompanied the work with a long and informative letter on performing practice. Haydn composed the Italian oratorio *Il ritorno di Tobia* (1774–5) for the annual Lenten concert of the Tonkünstler-Societät in Vienna, a charitable organization for musicians' widows and children founded by Hofkapellmeister Gassmann in 1771. He conducted the premières on 2 and 4 April 1775; most of the roles were sung by members of his own Kapelle. The work was a notable success; a review praised the choruses in particular and referred to his growing international reputation.

Beginning in 1766, the prince began to require operatic productions at the new castle; eventually opera would become the focus of the entire musical establishment. For the time being, however, Haydn's primary task was to compose operas to be produced during the festivities celebrating visits by high personages. Three comic operas date from the late 1760s: *La canterina* (1766) apparently had its première in the summer during a visit of the imperial court to Eisenstadt (in a makeshift theatre) and was afterwards produced in Pressburg (Bratislava). *Lo speziale* (1768) and *Le pescatrici* (1769–70) are both based on librettos by Goldoni; the former inaugurated the new opera house at Eszterháza probably during the last week of September 1768, on the visit of the Hungarian regent, Duke Albert of Saxe-Techsen, while the latter had its première on 16 September 1770 during the wedding celebrations of Countess Lamberg, the prince's niece. After a pause, operatic composition resumed in 1773 with *L'infedeltà delusa*, given on 26 July (the nameday of the dowager Princess Maria Anna, Paul Anton's widow), and *Philemon und Baucis, oder Jupiters Reise auf der Erde*, a German marionette opera, given on 2 September during the festivities in honour of a 'state' visit by Empress Maria Theresa to Eszterháza. (*Hexenschabbas*, another marionette opera from about this time, is lost.) *L'infedeltà delusa* was also given for the Empress; the performance occasioned her famous remark (if it is genuine) that in order to see a good opera she had to go to the country. Haydn's last opera during this phase was *L'incontro improvviso*, first given on 29 August 1775, during a visit by Archduke Ferdinand and his court.

During the late 1760s and early 1770s Haydn continued to compose instrumental works, albeit at a slower rate than before (except during the operatic hiatus of 1770–72). But they became longer, more passionate and more daring. The symphonies comprise nos.26, 35, 38, 41–9, 52, 58–9, 65; many of these are among his best-known before the London period, as is evident from their nicknames: 'Lamentatione', 'Maria Theresia', 'La passione', 'Mourning', 'Farewell' etc. He also took up the string quartet, not cultivated since the 1750s, producing three increasingly imposing *opera* in rapid succession: op.9 (*c*1769–70), op.17 (1771) and op.20 (1772). The reason is unknown: there is no documentation of quartet performances at the Esterházy court, and it has been speculated that he composed them for Viennese patrons (Burney described the audience's transports at a performance of Haydn quartets in Vienna in September 1772). He also composed numerous keyboard sonatas for connoisseurs: H XVI:45 (1766), 19 (1767), 46 (late 1760s), 20 (in C minor, 1771), as well as seven lost works and one that survives only as a fragment (H XVI:2a–g, XIV:5) which, to judge from the incipits, were on the same scale. A few concertos date from this period as well. Many of these works are so bold and expressive that in this century they have been subsumed under the appellation *Sturm und Drang*. The term has been criticized: taken from the title of a play of 1776 by Maximilian Klinger, it properly pertains to a literary movement of the middle and late 1770s rather than a musical one of about 1768–72, and early proponents of this interpretation assumed implausibly and without evidence that these works expressed a 'romantic crisis' in Haydn's life. Nevertheless, his style during these years was distinctive; furthermore, similar traits are found in the contemporary music of many other Austrian composers, including the teenaged Mozart's G minor Symphony K183/173dB and D minor String Quartet K173.

In Haydn's case this development may have been related to his turn to vocal music beginning in 1766: perhaps the demand for expressive depth in sacred works and dramatic effectiveness in opera, as well as the tendency towards through-composition in both genres, stimulated this expansion of his instrumental music. In 1769 Nicolaus began engaging a theatrical troupe each summer season; from 1772 to 1777 it was the famous one directed by Carl Wahr, which played primarily comedies and other entertainments, although Shakespeare's tragedies were also mounted. It has been speculated that Haydn supplied incidental music for these productions (including even *Hamlet* and *King Lear*) and that some *Sturm und Drang* symphonies recycle this music, although the only documented example is Symphony no.60 ('Il distratto', 1774), from a very un-Shakespearean French comedy. In any case, from about 1773 Haydn's instrumental music became generally lighter in style – the reason (if any) is again unknown; there is no evidence of princely intervention – and was again addressed to amateurs as well as connoisseurs. The string quartet was abandoned. Both the symphonies of 1773–5 (nos.50–51, 54–7, 60, 64, 66–9) and two contemporaneous sets of keyboard sonatas, H XVI:21–6 (1773) and especially 27–32 (1774–6), exemplify this mixed orientation; the former was published in Vienna in 1774 (the first authorized publication of Haydn's music) with a dedication to the prince, who presumably paid the costs. A third set (nos.35–9 and 20), again mixed in style, was published in 1780.

Opera impresario, 1776–90

In 1778 Haydn sold his house in Eisenstadt; the court now stayed at Eszterháza at least ten months every year, and he increasingly spent the short winter season in Vienna. The very long stays at Eszterháza were linked to Nicolaus's reorganization of the theatrical entertainment there in 1776. Now there was a regular 'season' each year, comprising opera, stage plays and marionette operas (in a separate small theatre); in principle there was theatrical entertainment every evening the prince was in residence. At first, stage plays predominated (184 evenings in 1778, as opposed to only 50 operas – and only two musical academies; four others took place during the day, in the 'apartments'), but the number of opera evenings increased steadily, reaching a high of 124 or 125 in 1786. New productions were henceforth not grand, 'occasional' events, but a regular occurrence; in 1776 there were five, and in the banner year 1786 there were eight, together with nine revivals. Under these conditions Haydn could not compose more than a small fraction what was needed, nor were new works commissioned from other composers. Instead, operas were acquired from the Veneto and from Vienna, where there was a lively trade in copying; it is not known how many were selected by the prince or Haydn during their brief winter sojourns. Others were acquired from agents (e.g. Nunziato Porta, the librettist of *Orlando paladino*), and newly arrived singers etc. and still others purchased from archives and estates (Dittersdorf sold the court several of his own operas in 1776). The up-to-date repertory centred around *opera buffa*: the composer represented by the greatest number of productions from 1776 to 1790 was Cimarosa (13), followed in order by Anfossi, Paisiello, Sarti and Haydn (seven), and 24 other composers with fewer.

19

Once it was decided to produce a given opera, Haydn was responsible for any musical alterations that might be required, supervising the copying of parts, rehearsing the singers and orchestra, and conducting all the performances – for no fewer than 88 productions in the 15 years from 1776 to 1790. This was by any reckoning a full-time job, even if one does not count his own new stage works, of which six originated between 1777 and 1783, or almost one per year. First came a *dramma giocoso* by Goldoni, *Il mondo della luna* (given on 3 August 1777, on the marriage of Nicolaus's younger son). *La vera costanza* (1778–9) is the subject of implausible and conflicting anecdotes in Griesinger and Dies, according to which it was originally commissioned for the Burgtheater in Vienna but scuttled by intrigue (neither Joseph II nor his musicians were well-disposed towards Haydn); in fact it had its première at Eszterháza, on 25 April 1779. It was lost in the fire that largely destroyed the Eszterháza opera theatre on 18 November 1779; the surviving version represents Haydn's reconstruction of the work from 1785. It is a measure of the prince's commitment (or obsession) that an opera was given just three days after the fire, in the marionette theatre, which had been hastily adapted for staged opera (yet another noble marriage was to be celebrated). Haydn's *L'isola disabitata* to a libretto by Metastasio also had its première on schedule on 6 December (the prince's nameday). Next came *La fedeltà premiata* (1780; given 25 February 1781, on the inauguration of the rebuilt opera house). In 1783 Haydn took the unusual step of publishing the great *scena* for Celia in Act 2, 'Ah, come il core ... Ombra del caro bene', in full score; it received a detailed and laudatory review by C.F. Cramer in his *Magazin der Musik*. Haydn's last two Eszterháza operas were *Orlando paladino* (1782, for the prince's nameday) and *Armida* (1783; given 26 February 1784). The later 1770s saw three German marionette operas, also all lost in the 1779 fire: *Dido* (1776), *Vom abgebrannten Haus* (date uncertain) and *Die bestrafte Rachbegierde* (1779; its production can be inferred only from the printed libretto); the occasionally seen *Die Feuersbrunst* is either spurious or represents an arrangement of *Vom abgebrannten Haus*.

After 1783 Haydn composed no more operas for the court. It is not known why he abandoned the genre, which he had cultivated intensively since 1766 and in which he was proud of his achievements, or how he persuaded the prince to consent at a time when the number of productions was still rising. Perhaps he was increasingly drawn to his new career as composer of instrumental music for publication. In any case all his other duties for the court theatre remained in force; in particular he still revised the operas in production to suit his provincial stage and limited personnel. Haydn made many cuts, both of entire numbers and within them, re-orchestrated (often adding winds), changed tempos (usually speeding them up) and 'tailored' arias to 'fit' his singers, as Mozart would have said. He composed about 20 substitute ('insertion') arias (H XXIVb) as well as long passages within numbers not rejected as a whole.

The majority of the insertion arias and simplifications were composed for Luigia Polzelli, a young Italian mezzo-soprano who joined the troupe in March 1779 along with her much older husband, a violinist. Both proved inadequate and were dismissed in December 1780 – but promptly rehired: Luigia and Haydn had become lovers, a relationship that, like so many in that milieu, was probably an open secret. (Haydn told Griesinger that the painter Ludwig Guttenbrunn had been his wife's lover during his stay at the court, 1770–72.) While at Eszterháza Luigia gave birth to her second son, Antonio, in 1783. He and his mother believed that Haydn was the father (there is no evidence of such a belief on Haydn's part); he became a professional musician and was appointed to the Esterházy orchestra in 1803. Haydn was well-disposed towards him, and even more towards his older brother Pietro (*b*1777); he taught them both music and maintained contact with them throughout his life. As for Luigia, following the dissolution of the Kapelle in 1790 Haydn attempted to procure engagements for her in Italy; however, he would not have her with him in London, even though her sister was engaged there as a singer. Although there are no letters from the 1780s by which we might assess the nature of their feelings, he wrote to her often (in Italian) during his first London visit. Those through early spring 1792 are ardent: 'Perhaps I shall never regain the good humour that I used to have with you; you are always in my heart, and I shall never, never forget you ... Think from time to time about your Haydn, who esteems you and loves you tenderly, and will always be faithful to you' (14 January). But those from May and June are notably cooler – he had entered a new relationship – and none survive from his second London visit. He acceded to Polzelli's requests for money, but not always immediately or in the demanded amount, while complaining (misleadingly) how little he had, as well as (accurately) how hard he had to work.

The vastly increased operatic and theatrical activity at the Esterházy court from 1776 on led to an equally drastic reduction in the performance of instrumental music. As noted above, only six 'academies' were listed for the entire year 1778 (all in January and February). Presumably the prince simply lost interest; even Haydn's stream of baryton works began to dry up after 1773 and ceased entirely about 1775, following the octets H X:1–6. The symphony, from the late 1750s up to 1775 the one constant in Haydn's output, declined as well; only nine were completed in the six years 1776–81 (nos.53, 61–3, 70–1, 73–5), none at all in 1777 or the first half of 1778. Even these few symphonies often include adaptations of stage music. No.63 in C begins with the overture to *Il mondo della luna*, and the slow movement ('La Roxelane') may be based on a theme from a stage play; the slow movement of no.73, 'La chasse', uses his own lied *Gegenliebe* and the finale recycles the overture to *La fedeltà premiata*. He even recycled the overture H Ia:7 twice, in the finale of one version of no.53 ('Imperial') and the opening movement of no.62. From this time on, the Esterázy court was no longer the primary destination for Haydn's instrumental music.

INDEPENDENCE, 1779–90

Nevertheless, Haydn was able to continue his career as an instrumental composer. In contrast to London, Paris and elsewhere, where unauthorized editions of his music had been appearing steadily since 1764, there was no music publishing industry to speak of in the Habsburg realm; most music circulated in manuscript copies. This situation changed in 1778, when Artaria & Co., hitherto primarily art dealers and mapmakers, expanded into music printing; other firms soon followed. Artaria and Haydn must have made contact in 1779 (it is not known who took the initiative); their first publication was a set of six keyboard sonatas, H XVI:35–9, 20 (delivered in the winter of 1779–80, published in April 1780), dedicated to the virtuoso sisters Katharina and Marianna von Auenbrugger. Dozens of Viennese publications of Haydn's music followed over the next decade. This would not have been possible on the terms of his 1761 contract, which forbade him from selling music on his own or composing for anyone else without permission. However, he signed a new contract on New Year's Day 1779, in which these prohibitions were omitted; the conjunction with Artaria's founding in 1778 and Haydn's publication of music with them beginning in 1779 or 1780 cannot be coincidental. The prince was losing interest in instrumental music; Haydn must have persuaded him to strike a compromise, whereby he remained in residence at court, continued in charge of the opera and drew his full salary, but was granted compositional independence in other respects, including the income from sales of his music. In addition, he began to market his music in other countries: in England beginning in 1781 with Forster, to whom he sold more music than to anyone except Artaria; in France beginning in 1783, selling Symphonies nos.76–8 (composed 1782) to Boyer and offering nos.79–81 (1783–4) to him as well (not to Naderman). (To be sure, certain works not composed for the court – for example, the 'Paris' Symphonies – were still performed, or at any rate tried out, there before being sent into the world, and others, such as the pino sonatas HXVI:40–42, were dedicated to members of the princely family.)

Haydn soon learnt to maximize his income by selling a given work in several countries, accepting a separate fee for each. Except in Vienna and London he often worked through a middleman. These activities were in many respects unregulated (modern copyright law being in its infancy); unauthorized 'double copying' was a constant danger, and everyone attempted to maximize his advantage – including Haydn, whose tactics were often unscrupulous, to say the least. He often earned his 'little extra' by selling manuscript copies of new works to private individuals; such 'subscription' copies still carried a certain prestige. An example is offered by his famous letters offering the string quartets op.33, composed in the summer and autumn of 1781 and sold to Artaria by prior arrangement. On 3 December he wrote to between ten and 20 noble and well-to-do music lovers, including the Swiss intellectual Johann Caspar Lavater:

> I love and happily read your works ... Since I know that in Zürich and Winterthur there are many gentlemen amateurs and great connoisseurs and patrons of music, I cannot conceal from you the fact that I am issuing a work consisting of 6 Quartets for two violins, viola and violoncello concertante, by subscription for the price of six ducats; they are of a new and entirely special kind, for I haven't written any for ten years ... Subscribers who live abroad will receive them before I issue the works here ...

However, Artaria (who presumably knew nothing of these activities) announced the forthcoming publication of the quartets on 29 December at a price of four gulden (six ducats equalled approximately 25 gulden). Haydn was furious:

> It was with astonishment that I read ... that you intend to publish my quartets in four weeks ... Such a proceeding places me in a most dishonourable position and is very damaging; it is a most extortionate step on your part ... Mr Hummel [the publisher] also wanted to be a subscriber, but I did not want to behave so shabbily, and I did not send them to Berlin solely out of regard for our friendship and further transactions; by God! you owe me more than 50 ducats, since I have not yet satisfied many of the subscribers, and cannot possibly send copies to those living abroad; this step must cause the cessation of further transactions between us.

In fact, it did not come to a rupture: Artaria delayed publication until April, and Haydn apparently sold the quartets to Hummel after all; both parties now better understood the ground rules ('the next time', wrote Haydn later, 'we shall both be more prudent'). A loss of 50 ducats implies about eight unsold copies; in 1784 Haydn claimed to Artaria that he had 'always received more than 100 ducats through subscriptions to my quartets'. Even as his publications increased, Haydn continued to market manuscript copies, especially in genres that were not ordinarily published (such as sacred vocal music), and to sell all sorts of music in places where there was still no music publishing industry, notably Spain. These were hardly ever new works. To be sure, he wrote to Artaria in 1784: 'The quartets I'm working on just now ... are very small and with only three movements; they are destined for Spain', but no trace of such works survives, unless it be the small-scale (but four-movement) single quartet op.42 (1785).

Another risk arose from the circumstance that many publishers sold works from their own catalogues to business partners in other markets. Forster naturally assumed he had exclusive rights in England to the works Haydn had sold him. However, when Artaria sold some of the same works to Longman & Broderip, two ostensibly authorized editions were suddenly in direct competition. To make matters worse, among the works Haydn sold Forster was a set of piano trios H XV:3–5, the first two of which were in fact compositions of his former student Pleyel. Later, Pleyel sold them to Longman & Broderip; when the latter edition appeared, Forster embarked on a lawsuit with Longman, in which Haydn became entangled when he went to England; it was settled out of court. Despite such difficulties, his methods of exploiting multiple markets became a model for the next two generations of composers; he 'taught' it to Beethoven (who learnt his lesson well, including the unscrupulous aspects), and it was still used by Mendelssohn and Chopin. He was also adept at 'marketing'. He described Symphonies nos.76–8 as 'beautiful, impressive and above all not very long symphonies ... and in particular everything very easy', and his first authorized Viennese publication of orchestral music (late 1782) was devoted, not to symphonies, but to the 'easier' genre of the overture.

For all these reasons Haydn's compositional activity underwent a sea-change in the 1780s. His music, which been well known and much praised since the mid-1760s, was now genuinely popular: he could scarcely keep up with the demand. He concentrated on what was saleable: instrumental works that would appeal to both amateurs and connoisseurs, opera excerpts and lieder. As long as his works had been destined for the court or published without his participation, he had had little need to follow the 'opus' principle; now he adopted it for almost all his publications. Even the string quartet was subject to another pause of six years (and the example of Mozart's quartets dedicated to him) before he composed three sets in rapid succession during 1787–90: op.50 (Artaria; dedicated to the King of Prussia), op.54/5 (a single set of six, sold to Johann Tost, formerly a violinist at court, who resold them to various publishers) and op.64. The English publisher John Bland visited him at Eszterháza in November 1789, when Haydn promised him a new quartet in return for a new razor (Haydn thanked him for the razors in April 1790). However, a 'new' quartet could not have been the one now known as the 'Razor' (op.55 no.2, composed in 1788 and never published by Bland); it is more likely that the story has to do with op.64 (1790), which Bland did publish in an authorized edition.

A genre that Haydn had not cultivated since the mid-1760s but which now again became important was the piano trio, with 13 works in the 1780s. HXV:5 (1784) and 9–10 (1785) were sold to Forster; nos.6–8 (1784–5) and 11–13 (1788–9) to Artaria, as was no.14 (1789–90). Nos.15–17 (1790) were composed for Bland; they specify a flute rather than a violin as the melody instrument (no.17: flute or violin). The piano sonatas HXVI:33, 34 and 43 were assembled *post facto* and published in 1783; by contrast, nos.40–42 are an 'opus' (published 1784; dedicated to Maria Hermenegild, wife of the later Prince Nicolaus II). Two important single sonatas date from 1789–90: no.48 in C, composed for Breitkopf in Leipzig, and no.49 in E♭, for Maria Anna von Genzinger; the Fantasia (Capriccio) HXVII:4, composed 'in a *launige* hour' in 1789, is equally fine. Another genre made newly popular through publication was the lied; Haydn composed 24 in 1781–4 (HXXVIa:1–24) and published them with Artaria in two sets of 12.

Even in the early 1780s Haydn was no mere 'entertainer'. But a newly serious orientation was instigated by his receipt in about 1784–5 of two prestigious commissions from abroad, both executed 1785–6. Six symphonies were commissioned by Count d'Ogny for performance in Paris by the Concert de la Loge Olympique (a Masonic organization); the fee was later reported to have been 25 louis d'or for each symphony (Mozart had been paid only five for the 'Paris' Symphony KV297), plus five louis d'or for the publication rights (Imbault, 1788). The Concert employed a much larger orchestra than any for which he had composed symphonies; whether for this reason or simply owing to the notion of 'Paris', they are the grandest he had yet written. They were immensely popular; Marie

25

Antoinette supposedly preferred no.85 in B♭, whence its nickname 'La reine'; compare 'L'ours' (no.82) and 'La poule' (no.83). Their success led to additional symphonies: Haydn sold nos.88–9 (1787) to Tost, who resold them in Paris and elsewhere, and d'Ogny commissioned nos.90–92 (1788–9).

The other commission was a highly unusual one from Cádiz, for a series of orchestral pieces on the last words of Christ, to be performed in a darkened church as a kind of Passion during Holy Week, presumably on Good Friday. Haydn described them to Forster as

> purely instrumental music divided into seven Sonatas, each Sonata lasting seven or eight minutes, together with an opening Introduction and concluding with a *Terremoto* or Earthquake. These Sonatas are composed on, and appropriate to, the Words that Christ our Saviour spoke on the Cross. ...
>
> Each Sonata, or rather each setting of the text, is expressed only by instrumental music, but in such a way that it creates the most profound impression on even the most inexperienced listener.

Griesinger commented: 'Haydn often stated that this work was one of his most successful'. It was widely performed and favourably received, not least owing to its avoidance of what were taken to be the chief dangers of tone-painting, excessive literalness and triviality. Haydn also sold the *Seven Last Words* in arrangements, one for string quartet and one for keyboard.

A distinctly lighter series of commissions came from King Ferdinando IV of Naples. Like Prince Esterházy, he had become proficient on an out-of-the-way instrument: the *lira organizzata*, a sort of grown-up hurdy-gurdy. He commissioned concertos and 'notturnos' for two *lire organizzate* from various composers; Haydn supplied five or six concertos (HVIIh, 1786–7) and eight notturnos (HII:25–32, 1789–90).

Haydn's stays in Vienna were still restricted to one or two months each winter and occasional brief visits during Lent. He increasingly valued the imperial capital's artistic and intellectual life, which was flourishing under Joseph II, and chafed at having to spend so much time in the 'wasteland' (*Einöde*) of Eszterháza. He acquired many friends and patrons, including Baron Gottfried van Swieten (whom he had met in 1775), Councillor Franz Sales von Greiner (1730–98; father of the later Caroline Pichler), who presided over Vienna's leading literary salon and supplied Haydn with lieder texts, Anton Liebe von Kreutzner, who in 1781–2 commissioned the *Mariazellermesse* and to whose daughter Haydn dedicated his lieder published in 1781, Councillor Franz Bernhard von Keess, who held regular concerts of orchestral music and made the first systematic attempt to collect all of Haydn's symphonies, and Michael Puchberg (Mozart's patron). Their orientation, reflecting that of the emperor, was enlightened-conservative: they were interested in literature, philosophy and education and largely rejected dogmatism, yet retained a traditional and Catholic outlook – all traits that Haydn shared. The majority were freemasons; Nicolaus himself was Master of Ceremonies at one Viennese lodge, and it was most probably he or others in this circle who induced Haydn to apply for membership in the order. Haydn did so on 29 December 1784 and was inducted into the lodge 'Zur wahren Eintracht' on 11 February 1785; however, there are no further records of his participation, and (despite one further letter) it appears that freemasonry was of no particular significance to him.

Haydn's visits to Vienna offered many opportunities for performances of his music. The string quartets op.33 are still sometimes called the 'Russian' quartets, owing to a dedication on a late edition that reflects a performance given on Christmas Day 1781 for Grand Duke Paul of Russia (later Tsar Paul I) and his music-loving consort. Regarding his lieder, Haydn told Artaria: 'I will sing them myself, in the best houses. A master must see to his rights by his presence and by correct performance'. In February 1779 the Tonkünstler-Societät had invited Haydn to join, but attached conditions not to his liking; he gruffly refused. In March 1784, however, he produced *Il ritorno di Tobia* for them in a revised version. The 'tightness' of the Viennese performing scene is evident from the fact that among the five soloists were four who had taken part (or would do so) in Mozart opera premières:

Anna	Nancy Storace (*Figaro*, Susanna)
Raffaelle	Catarina Cavalieri (*Entführung*, Konstanze)
Tobia	Valentin Adamberger (*Entführung*, Belmonte)
Tobit	Stefano Mandini (*Figaro*, Count Almaviva)

Haydn and Storace became warm friends; he later composed a cantata 'for the voice of my dear Storace' (possibly *Miseri noi*, H XXIVa:7). In January 1787 three of the 'Paris' symphonies were performed, and in March the *Seven Last Words* at the Palais Auersperg; both were unpublished novelties at the time.

The friendship between Haydn and Mozart also developed in Vienna. It is believed that they first met in 1783 or 1784, at a performance such as that of *Tobia* just described, one of Mozart's 'academies', or at a quartet party: Michael Kelly's (late and perhaps untrustworthy) reminiscence of Stephen Storace's quartet comprising Haydn, Dittersdorf, Mozart and Vanhal is set in 1784 (Kelly later visited Haydn at the Esterházy court). Mozart performed his six new quartets for 'my dear friend Haydn and other good friends' on 15 January 1785, and the last three again on 12 February; the latter occasioned Haydn's famous remark to Leopold Mozart that Wolfgang was the greatest composer he knew, 'either by name or reputation'. And Mozart's dedicatory letter in Artaria's edition of the quartets (September 1785) is headed: 'To my dear friend Haydn'. In the winter of 1789–90 he invited Haydn to rehearsals of *Così fan tutte*, and Haydn organized a quartet party in which Mozart's participation can be inferred.

The nature of their relationship has been much discussed. Many writers have romanticized it, beginning with Griesinger's and Dies's sentimental accounts of their tearful farewell in December 1790 on Haydn's departure for London (including Haydn's alleged comment, 'My language is understood in the entire world'). Others are sceptical, noting that the surviving documentation derives solely from two winters, 1784–5 and 1789–90, and that Mozart's dedicatory letter may protest his friendship a little too much. But there is no doubt of their mutual admiration as composers: each acknowledged the other as his only peer and as the only meaningful influence on his own music in the 1780s. Mozart's dedication of quartets to Haydn – a mere composer rather than a rich or noble patron – was unusual (although cynics note that he might have attempted to recruit the latter, but failed), and Griesinger relates an anecdote according to which he defended Haydn against a stupid criticism by Kozeluch: 'neither you nor I would have hit on that idea'. But Haydn's expressions of admiration went further. In a famous (albeit unauthenticated) letter of 1787 to the impresario Franz Rott in Prague, he admitted that he feared comparison with 'the great Mozart', at least on the stage:

> If only I could impress Mozart's inimitable works on the soul of every friend of music, and the souls of high personages in particular, as deeply, with the same musical understanding and with the same deep feeling, as I understand and feel them, the nations would vie with each other to possess such a jewel.

In early 1792 he wrote to Puchberg: 'I was quite beside myself for some time over [Mozart's] death and could not believe that Providence would transport so irreplaceable a man to the other world', adding that he had offered to teach Mozart's son Karl without fee (something he did not do lightly). 'I have often been flattered by my friends with having some genius', he said in Burney's hearing, 'but he was much my superior'. It is remarkable that his feelings were apparently marked neither by jealousy nor a compromise of his musical self-confidence, except possibly regarding opera; they had no effect on his productivity.

In any case it was a friend of a different sort whom Haydn most cherished around 1790: Maria Anna von Genzinger (1750–93), the wife of an important physician (whose clients included Nicolaus Esterházy) and a talented amateur pianist. In June 1789 she sent Haydn her piano arrangement of the slow movement of an unidentified composition; he responded with praise, and the relationship rapidly became intense, although as far as can be told it remained platonic. Haydn's letters to Mme Genzinger are his most fervent and intimate; that of 9 February 1790, following his sudden return from Vienna to Eszterháza, is at once poignant and amusing:

> Here I sit in my wilderness – forsaken – like a poor waif – almost without human society – sad – full of the memories of past glorious days – yes! past, alas! – and who knows if those days will return again? Those wonderful parties? – where the whole circle is one heart, one soul – all the beautiful musical evenings? ... For three days I didn't know if I was Kapellmeister or Kapell-servant. Nothing could console me, my whole house was in confusion, my pianoforte, which I usually love, was perverse and disobedient ... I could sleep only a little, even my dreams persecuted me; and then, just when I was happily dreaming that I was listening to *Le nozze di Figaro*, the horrible North wind woke me and almost blew my nightcap off my head ... Alas! alas! I thought to myself as I was eating here, instead of that delicious slice of beef, a chunk of a 50-year-old cow ... Here in Eszterháza no one asks me: 'Would you like some chocolate, with milk or without? ... What may I offer you, my dear Haydn, would you like a vanilla or a strawberry ice?'

But, as always, he soon recovered; the letter continues: 'I am gradually getting used to country life, and yesterday I composed [*studierte*] for the first time, and indeed quite Haydnish'. Later that year he composed a sonata (H XVI:49) for Mme Genzinger, in the course of which he also advised her on the purchase of a new fortepiano; earlier he had advised her daughter Josepha about her performances of his cantata *Arianna a Naxos* (H XXVIb:2).

But the year 1790 was to prove even more disruptive than Haydn could have suspected in early February. Joseph II died on 20 February, throwing Vienna into mourning; five days later, Nicolaus Esterházy's wife died (Haydn had his hands full keeping him from succumbing to depression), followed on 28 September by the prince himself. Anton, his son and successor, immediately dissolved the musical and theatrical establishment, although Haydn was kept on at a reduced salary without official duties; he also received 1000 gulden a year from Nicolaus's estate.

He at once moved to Vienna, taking rooms with a friend, J.N. Hamberger. He declined an offer to become Kapellmeister for Prince Grassalkovics (Nicolaus's son-in-law, resident in Pressburg), and made it clear to King Ferdinando that he would not fulfil any vague promises he might have made to travel to Naples. Whatever his intentions were, they were soon overtaken by events.

4. LONDON, 1791–5

Johann Peter Salomon, born in Bonn but living in London as violinist and concert producer, heard of Nicolaus Esterházy's death while in Europe engaging musicians for the coming season; he immediately travelled to Vienna, called on Haydn and 'informed' him that he would now be going to London. Salomon was not the first to contemplate this. In 1783 Haydn said of Symphonies nos.76–8 that they were composed 'for the English gentlemen, and I intended to bring them over myself and produce them' at the Professional Concerts (the successor to the Bach-Abel Concerts of 1774–82); in July 1787 he contemplated composing an opera and instrumental works for G.A.B. Gallini, the impresario of the Italian opera. Salomon himself had had business dealings with Haydn, who in a letter to Bland of April 1790 referred to 40 ducats owed him by Salomon. However, as long as Nicolaus Esterházy was alive Haydn had been unwilling or unable to negotiate his freedom; now, Prince Anton willingly granted him a year's leave.

Salomon's initial contract with Haydn governed the 1791 season. Haydn was guaranteed £300 for an opera (here Salomon was acting as agent for Gallini), £300 for six symphonies, £200 for the rights to publish the latter, £200 for 20 other compositions to be conducted at his concerts and £200 profit from a 'benefit' concert. Haydn and Salomon left Vienna on 15 December, travelling via Munich, Wallerstein (where Haydn conducted Symphony no.92) and Bonn to Calais, from where they sailed to Dover on New Year's Day 1791, arriving in London the next day. 'I stayed on deck during the entire crossing', he wrote to Mme Genzinger on 8 January,' so as to gaze my fill of that great monster, the ocean'.

London was the largest and economically most vibrant city in the world, made even more cosmopolitan by refugees from the French Revolution. Haydn settled in the same house where Salomon lived and also had a studio at Broadwood's music shop, although he complained of the noise and later moved to a suburb. He immediately plunged into a hectic social and professional life which, however stimulating, competed with his need to compose: 'My arrival caused a great sensation ... I went the rounds of all the newspapers for three successive days. Everyone wants to know me ... If I wanted, I could dine out every day; but first I must consider my health, and second my work. Except for the nobility, I admit no callers until 2 o'clock.' He also had to give fortepiano lessons, primarily to high personages; he was soon invited to a ball at the Court of St James's, and to a concert sponsored by the Prince of Wales (later George IV). The latter was an enthusiastic amateur who became Haydn's most important royal patron; on 1 February 1795 he arranged a soirée with George III and the Queen in attendance, in which all the music was of Haydn's composition and he both played and sang. Other royal invitations followed that month and in April.

London's musical life was active and varied, and enriched by a constant stream of artists from abroad. It is abundantly documented; about Haydn's activities there are also his letters to Genzinger and Polzelli, as well as the observations and anecdotes jotted down in his 'London Notebooks'. The 'season' ran from February to May; it included two series of concerts in the Hanover Square Rooms (Salomon and Haydn on Monday and the Professional Concert on Friday), opera on Tuesday and Saturday, 'ancient music' on Wednesday etc. Performing forces were generally larger than in Vienna or Eszterháza; in the 1791–2 season Haydn's symphonies were performed by about 40 players, in 1795 about 60. The typical concert was a mixed affair, including symphonies, sonatas, arias and duets. A special feature was the massed choral performances in Westminster Abbey; Haydn's experience of hearing Handel's oratorios there was the chief stimulus for *The Creation*.

Haydn arrived with relatively few new works except the string quartets op.64; they were published in 1791–2 by Bland, 'composed by

Giuseppe Haydn, and performed under his direction, at Mr Salomon's Concert'. However, Symphonies nos.90–92 had not yet been printed in England; Haydn made use of nos.90 and 92 in 1791, as well as lyre notturnos (arranged for flute and oboe) and *Arianna a Naxos*. Symphony no.92 soon became a favourite, and was one of several symphonies performed on the occasion of his receiving the honorary doctorate of music at Oxford, 6–8 July 1791 (whence its nickname). This event meant a great deal to him; thereafter he often referred to himself in public documents (or when needing to assert his status) as 'Doktor der Tonkunst'.

Among Haydn's new compositions in 1791–2 were the first six 'London' symphonies (nos.93–8). Only nos.95 in C minor and 96 in D ('Miracle') were given in 1791; the others followed in 1792, although nos.93–4 had been composed in the second half of 1791: no.93 in D (17 February), 98 in Bb (2 March), 94 in G ('Surprise', 23 March), 97 in C (3 or 4 May). His benefit concert in May cleared £350, nearly double the £200 he had been guaranteed. Even though the journalistic prose is somewhat formulaic and may to some extent even have been 'suggested' by Salomon, these concerts were a sensation. The following report is of the first, 11 March 1791:

> Never, perhaps, was there a richer musical treat. It is not wonderful that to souls capable of being touched by music, HAYDN should be an object of homage, and even of idolatry; for like our own SHAKSPEARE he moves and governs the passions at will. His new *Grand Overture* [symphony] was pronounced by every scientific ear to be a most wonderful composition; but the first movement in particular rises in grandeur of the subject, and in the rich variety of air and passion, beyond any even of his own productions.

The symphony was no.92, 95 or 96. Charles Burney, with whom Haydn was to become fast friends, waxed equally enthusiastic:

> Haydn himself presided at the piano-forte; and the sight of that renowned composer so electrified the audience, as to excite an attention and a pleasure superior to any that had ever, to my knowledge, been caused by instrumental music in England. All the slow middle movements were encored; which never happened before, I believe, in any country.

By contrast, Haydn's new opera, *L'anima del filosofo* or *Orfeo ed Euridice*, was never produced: although he composed it during the spring of 1791 and it had entered rehearsal, Gallini (owing to political intrigue) was denied a licence for the theatre.

By the end of the 1791 season Salomon and Haydn were agreed that he would stay for another year. However, Anton Esterházy (who had written to Haydn cordially in February) wanted him to return; when Haydn informed him that he had signed a new contract with Salomon and requested an additional year's leave, the prince refused, demanding that he inform him 'by the next post the exact time when you will arrive back here again'. Haydn feared outright dismissal; as so often, it did not come to that. The summer and autumn gave him ample opportunity to compose, travel and expand his social circle. In August and early September he stayed on the estate of Nathaniel Brassey, a banker, in Hertfordshire. By mid-September he was back in London; on 5 November he attended an official banquet given by the newly-installed Lord Mayor (amusingly described in the Notebooks). On 24–5 November he stayed at Oatlands, a property of the Duke of York, who the previous day had been married to a daughter of the King of Prussia; the new duchess became one of his most loyal patrons.

Meanwhile, he had begun giving piano lessons to Rebecca Schroeter, the attractive and well-to-do widow of the composer and pianist Johann Samuel Schroeter. During the winter of 1791–2 this relationship blossomed into a passionate affair, documented by copies Haydn made of her letters to him (beginning with this one, from 7 March):

> My D[ear]: I was extremely sorry to part with you so suddenly last Night ... I had a thousand affectionate things to say to you, my heart was and is full of TENDERNESS for you, you are DEARER to me every day of my life ... I am truly sensible of your goodness, and I assure you my D. if anything had happened to trouble me, I would have opened my heart, & told you with the most perfect confidence. Oh how earnestly [I] wish to see you. I hope you will come to me tomorrow. I shall be happy to see you both in the Morning and the Evening....

Haydn later confirmed the relationship to Dies: 'an English widow in London, who loved me ... a beautiful and charming woman and I would have married her very easily if I had been free'. Although no letters between them survive following his departure in the summer of 1792, they remained close: Haydn dedicated the piano trios HXV:24–6 (1795) to her, she witnessed an important contract of 1796 regarding English editions of his music, they were in touch again regarding business in 1797 and she was a subscriber to *The Creation*.

Haydn's and Salomon's plans for the 1792 season were complicated by the organizers of the Professional Concert, whom Haydn had disappointed in the 1780s and who in 1791 had printed scurrilous notices alleging that his talent had dried up. As he wrote to Mme Genzinger (17 January), complaining of overwork:

> At present I am [composing] for Salomon's concerts, and I am making every effort to do my best, because our rivals, the Professional Concert, have had my pupil Pleyel from Strasbourg come here to conduct their concerts. So now a bloody harmonious war will commence between master and pupil. The newspapers are all full of it, but it seems to me that there will soon be an armistice, because my reputation is so firmly established. Pleyel behaved so modestly towards me on his arrival that he won my affection again.

Pleyel's music, similar to but far less brilliant and complex than his master's, was then very popular. Haydn's outward modesty and good manners notwithstanding, he was not about to be upstaged by a former student whose talent lay far beneath his own. Even his most famous 'surprise', in the eponymous symphony, played a role. Griesinger writes:

> I jokingly asked him once if it was true that he composed the Andante ... in order to wake up the dozing English audience. 'No ... my intention was to surprise the public with something new, and to debut in a brilliant manner, in order to prevent my rank from being usurped by Pleyel, my pupil ...'

Another orchestral work of 1792 is the 'Concertante', a *symphonie concertante* for violin, cello, oboe and bassoon and orchestra (given on 9 March); it was composed in direct competition with Pleyel's most popular composition.

On 10 April 1792 Haydn wrote to Prince Anton, informing him that 'our concerts will be finished at the end of June, after which I shall begin the journey home without delay, in order to serve my most gracious prince and lord again'. He departed from London at the end of June or beginning of July, travelling via Bonn (where he met Beethoven) and Frankfurt. There is considerable evidence that he intended to return for the 1793 season; in the event he did not until 1794. On 24 July 1792 he arrived in Vienna, occupying the same lodgings as in the autumn of 1790. His 18 months in the Habsburg capital were uneventful. Polzelli was in Italy (still asking for money), and Mme Genzinger died on 20 January 1793. Haydn must have been lonely when not absorbed in his work, a condition presumably not ameliorated by his wife's suggestion that they purchase a small house in the suburb of Gumpendorf (not occupied until 1797; it is now the Vienna Haydn Museum). Beethoven arrived in November, and their intercourse began immediately; it included a stay at Eisenstadt with Haydn in the summer of 1793. Haydn set him (like all his students) to a systematic course of counterpoint on Fux's model, but had neither time nor inclination to correct the exercises systematically, and Beethoven switched to Albrechtsberger. More important to both composers was doubtless whatever comments Haydn made about free composition, as well as shop-talk about musical life and career-building (these seem to have been his chief contribution to the training of his many 'pupils' during the 1790s). By early winter he had decided not to return to London in 1793: he had finally determined to undergo an operation for a long-painful polyp in his nose, and perhaps he feared travelling at a dangerous stage in the Napoleonic Wars. In any case it was to his advantage to take an additional year, so as to be able to have new compositions in his portfolio on his return.

In November 1792 Haydn produced 12 each of new minuets and German dances (H IX:11–12) at a charity ball, and immediately published them with Artaria. Keess had produced Symphonies nos.95–6 in the winter of 1791–2; Haydn produced others himself on 15 March 1793, and again for the Tonkünstler-Societät in December, where he also gave his choral 'madrigal' *The Storm* (H XXIVa:8) in a revised version with a German text, possibly by Swieten. He began the string quartets op 71/74 (a single opus of six) in late 1792 and composed them mainly in 1793, with a view to producing them in London. He also refined his sales methods: a nobleman (in this case Count Anton Georg Apponyi) purchased the 'dedication' for 100 ducats, for which he received the exclusive right (in Vienna) to own a manuscript copy and to perform the quartets until Haydn published them (after one or two years), and was named as dedicatee on the editions. Later in 1793 Haydn worked on three symphonies for London, completing no.99 in E♭, the second and third movements of no.100 in G ('Military') and all but the first movement of no.101 in D ('Clock'). Another major composition from 1793 is the variations for piano in F minor, H XVII:6. On 19 January 1794 he departed for London, accompanied by the former Esterházy copyist Johann Elssler, now his amanuensis; they arrived on 5 February. On the way they stayed in Passau, where he heard a choral arrangement of the *Seven Last Words* by the local Kapellmeister, Joseph Friebert.

The Professional Concert having disbanded, in 1794 Haydn and Salomon had the stage to themselves. Symphony no.99 had its première on 10 February 1794, no.101 on 3 March, no.100 on 31 March. In the summer Haydn travelled to Portsmouth, the Isle of Wight, Bath and Bristol. For the 1795 season, however, Salomon abandoned his concerts, owing to the difficulty of obtaining 'vocal performers of the first rank from abroad' (he resumed them in 1796). Haydn therefore allied himself with the so-called Opera Concerts, directed by the violinist and composer Giovanni Battista Viotti, with an even larger orchestra, of approximately 60 players; Symphony no.102 was first given on 2 February, no.103 ('Drumroll') on 2 March, and no.104 – 'The 12th which I have composed in England', Haydn wrote on the autograph, doubtless with more than a touch of pride – at his benefit concert on 4 May. His success was greater than ever; following the benefit, Burney wrote that his 1795 symphonies were 'such as were never heard before, of any *mortal's* production; of what Apollo & the Muses compose or perform we can only judge by such productions as these'. Another important work given its première at this concert was the cantata *Berenice, che fai* (HXXIVa:10), composed for the reigning prima donna, Brigida Giorgi Banti.

Haydn's compositions from the period 1794–5 are more heterogeneous than those from 1791–2. He returned to piano music for the first time since 1790, composing at least three sets of trios: HXV:18–20, dedicated to Maria Therese Esterházy, the widow of Prince Anton; 21–3, dedicated to Maria Hermenegild, the wife of Anton's successor Prince Nicolaus II; and 24–6, dedicated to Mrs Schroeter. Nos.27–9 were dedicated to his friend Therese Jansen (*b* c1770), a celebrated virtuoso who in 1795 married a son of the engraver Francesco Bartolozzi (Haydn was a witness); it is unclear whether they date from 1795 or 1796. He composed no.31 for Jansen as well: the finale (1794) originated as an occasional piece titled 'Jacob's Dream!', designed to amuse her by showing up the insufficiencies of a self-important violinist in the higher registers. He also composed sonatas nos.50 and 52 for Jansen, and no.51 probably for Schroeter. Another genre he took up again owing to the influence of a lady was the solo song. The muse was Anne Hunter (1742–1821), another well-to-do widow (of the famous surgeon John Hunter) and a minor poet, who supplied the texts for at least nine of Haydn's 14 songs in English; 12 appeared as two sets of *Original Canzonettas* in 1794–5. He also composed numerous arias, divertimentos, marches, canons etc.

Haydn's London visits were the highpoint of his career up to that time. Griesinger reports that he earned 24,000 gulden and netted 13,000 (the equivalent of more than 20 years' salary at the Esterházy court), and that he 'considered the days spent in England the happiest of his life. He was everywhere appreciated there; it opened a new world to him'. Whether he seriously contemplated staying is not known; Prince Anton Esterházy had died in 1794, freeing him from even a nominal obligation to the court, and the royal family attempted to persuade him to remain. But the question was settled when Anton's successor, Nicolaus II, offered him reappointment as Esterházy Kapellmeister. Although he remained in London for two months following the end of the 1795 season, composing trios and canzonets and seeing to the publication of many of his English compositions – he established new, long-term relations, for example with the 'musick seller' F.A. Hyde, an agent for Longman & Broderip, with whom he signed an elaborate contract in 1796 – he departed (according to Dies) on 15 August, travelling via Hamburg and Dresden and arriving in Vienna presumably around the beginning of September. His new house still not being ready for occupancy, he took lodgings on the Neuer Markt in the old city.

5. VIENNA, 1795–1809

'Haydn often said that he first became famous in Germany owing to his reputation in England' (Griesinger). From 1761 to 1790, notwithstanding his fame as a composer and the brilliance of the Esterházy court, he had been 'stuck in the country' (letter to Artaria, 17 May 1781); he could never be in Vienna for very long, his music was not in favour at the imperial court and his relationship with the Tonkünstler was strained. In 1795, by contrast, he returned as a culture-hero. Many of his remaining works originated in collaboration with the cultural-political establishment and were staged as 'events' of social and ideological as well as musical import. The key figure was Baron van Swieten, the imperial librarian and censor and the resolutely high-minded leader of the Gesellschaft der Associirten, an organization of noble patrons who subsidized large-scale performances of oratorios and the like. In addition, Haydn's position as Esterházy Kapellmeister was far less onerous than before. Nicolaus II largely abandoned Eszterháza in favour of Vienna and Eisenstadt; Haydn's primary duty was to supply a mass each year to be performed in conjunction with the celebration of the nameday (8 September) of Maria Hermenegild, Nicolaus's consort; this took place in Eisenstadt, where he usually spent the summer.

As a result, Haydn's compositional orientation changed fundamentally. He composed little instrumental music: no orchestral works save the Trumpet Concerto (1796) and only one piano work (the trio no.30, 1796; nos.27–9 may have been completed then as well); the only genre he actively cultivated was the string quartet, with op.76 (completed 1797, dedicated to Count Joseph Erdődy; published 1799) and op.77 (1799, dedicated to Prince Joseph Franz Lobkowitz; published 1802). Instead, he devoted himself primarily to sacred vocal music: masses for Esterházy and oratorios for Vienna. Both the *Missa Sancti Bernardi von Offida* ('Heiligmesse') and the *Missa in tempore belli* ('Kriegsmesse'; 'Paukenmesse') are dated 1796; the former was apparently composed first and given its première in Eisenstadt, while the latter originated in the autumn and was first given at the Piaristenkirche in Vienna on 26 December, then produced in Eisenstadt in 1797. There followed the *Missa in angustiis* ('Nelson' Mass) in 1798, the 'Theresienmesse' in 1799, the 'Schöpfungsmesse' in 1801 and the 'Harmoniemesse' in 1802; thereafter this office was fulfilled by other composers, including Hummel and, in 1807, Beethoven. The absence of new masses in 1797 and 1800 doubtless reflects Haydn's intense work on *The Creation* and *The Seasons*, respectively, during those two years.

Haydn's collaboration with Swieten began in early 1796 with an arrangement of the *Seven Last Words* as a kind of oratorio; more precisely, a reworking of Friebert's arrangement. He added chorale-like *a cappella* intonations preceding most of the 'words' and a wind-band introduction to the second part, rewrote the choral parts and revised the orchestration; Swieten revised the text and arranged for the première at the Schwarzenberg Palace, 26–27 March. Haydn produced it often thereafter, usually for the benefit of the Tonkünstler-Societät. In 1797 that organization finally made up for its earlier neglect: on 20 January, a letter signed by Salieri and Paul Wranitzky granted Haydn free admission to all their concerts for life, and on 11 December he was elected a 'senior assessor' in perpetuity.

An overtly political composition was the 'Emperor's Hymn', *Gott erhalte Franz den Kaiser*, from the winter of 1796–7. Although one later account has Swieten manipulating things, the immediate impetus came from Count Joseph Franz Saurau, the president of Lower Austria and later Minister of the Interior:

> I have often regretted that unlike the English we had no national anthem suited to display before the entire world the devoted attachment of the people to their *Landesvater* ... This seemed especially necessary at a time when the Revolution in France was raging at its strongest ...

I had a text fashioned by the worthy poet [Lorenz Leopold] Haschka; and to have it set to music, I turned to our immortal compatriot Haydn, who, I felt, was the only man capable of creating something that could be placed at the side of ... 'God Save the King'.

Haydn identified thoroughly with the cultural politics of this project. In late January 1797 the hymn was hastily printed and disseminated, and performed in theatres throughout the Habsburg realm on the emperor's birthday, 12 February. This 'Volkslied', as he called it, combined hymnlike and popular elements so successfully that it became the anthem of both Austria and Germany. Later in 1797 he employed the melody as the basis for the variation movement in the String Quartet op.76 no.3, and in his last years he played it daily at the piano.

But Haydn and Swieten were already pursuing bigger game. On Haydn's departure from England Salomon had given him a libretto (now lost) entitled *The Creation of the World*, supposedly written for Handel but never set to music. Back in Vienna, Haydn showed it to Swieten: 'I recognized at once that such an exalted subject would give Haydn the opportunity *I had long desired*, to show the whole compass of his profound accomplishments and to express the full power of his inexhaustible genius'; the emphasized phrase implies that Swieten's purpose was not artistic alone, but cultural-political as well. The Associirten guaranteed Haydn 500 ducats and subsidized the copying and performance. Haydn, whose enthusiasm for the project was bound up with his experience of Handel in London, conceived the remarkable notion of disseminating the work in both German and English (it is apparently the first original bilingual composition). Swieten translated the libretto and adapted the English prosody to his German version; he also made suggestions regarding the musical setting, many of which Haydn adopted. He began the composition perhaps in autumn 1796 (an unauthenticated letter from Albrechtsberger to Beethoven states that he heard Haydn 'improvise' from it in December); it was at least half done by summer 1797 and (according to F.S. Silverstolpe) was completed during the autumn, in Eisenstadt, although the preparation of performance materials (which entailed revisions) lasted until March 1798.

The Associirten at first produced the work only in private, again at the Schwarzenberg Palace; following a 'Generalprobe' on 29 April the official première took place on the 30th, with additional performances on 7 and 10 May. The effect was overwhelming; Silverstolpe reported:

39

> No one, not even Baron van Swieten, had seen the page of the score wherein the Creation of Light is portrayed ... Haydn had the expression of someone who is thinking of biting his tongue, either to hide his embarrassment or to conceal a secret. And in that moment when Light broke forth for the first time, one would have said that light-rays darted from the composer's blazing eyes. The enchantment of the electrified Viennese was so profound that the performers could not proceed for some minutes.

The first public performance took place on 19 March 1799, at the Burgtheater, with a complement of about 180 performers (not 400, as has been speculated); a benefit for the Tonkünstler followed on 22 December. The work immediately became a staple especially of charity performances, many conducted by Haydn himself, which brought in tens of thousands of gulden. 'I know that God has favoured me', he said to Griesinger, 'but the world may as well know that I have been no useless member of society, and that one can also do good by means of music'. Beyond this, *The Creation* 'made history' immediately and on a pan-European scale in a way equalled by no other composition, owing to its fortunate combination of sublime subject, cultural-historical 'moment' on the cusp between Enlightenment and Romanticism, appeal to both high-minded and ordinary listeners, Haydn's unrivalled stature and the originality and grandeur of his music. His pride in and personal identification with the work, in addition to his usual concern for financial advantage, induced him to publish it himself, selling it 'by subscription' all over Europe, with the assistance of colleagues such as Dr Burney; his advertisement (June 1799) reads:

> The success which my Oratorio *The Creation* has been fortunate enough to enjoy ... [has] induced me to arrange for its dissemination myself. Thus the work will appear ... neatly and correctly engraved and printed on good paper, with German and English texts; and in full score, so that [at least] one work of my composition will be available to the public in its entirety, and the connoisseur will be in a position to see it as a whole and to judge it.

The edition appeared at the end of February 1800 with a list of more than 400 subscribers.

By spring 1799 Haydn and Swieten were planning a second oratorio, *The Seasons*, with a libretto based on James Thomson's pastoral epic of 1726–8; Haydn composed the music apparently from the autumn of 1799 to the end of 1800. He suffered a serious illness in the winter of 1800–1, as the work neared completion, during which he again identified with his own oratorio. His student Sigismund Ritter von Neukomm reported:

> Speaking of the penultimate aria, 'Behold, O weak and foolish man, Behold the picture of thy life' ... he said: 'This aria refers to *me!*' And in this wonderful masterpiece he really did speak entirely from his inmost soul, so much so that he became seriously ill while composing it, and ... the Lord ... allowed him to see 'his life's image and his open grave'.

The private première at the Schwarzenberg Palace took place on 24 April 1801, the first public production on 19 May. Although the initial reception of *The Seasons* was favourable – Haydn wrote to Clementi that it had enjoyed 'unanimous approval' and that 'many prefer it to *The Creation*, because of its greater variety' – critical opinion soon became mixed, owing in part to its perceived 'lower' subject, in part to a growing aesthetic resistance to its many pictorialisms. Haydn himself contributed to both strands of criticism: he supposedly said to Franz II, 'In *The Creation* angels speak and tell of God, but in *The Seasons* only Simon speaks' (Dies); and he indiscreetly criticized Swieten's croaking frogs ('Frenchified trash') and the absurdity of a choral hymn to toil (*Fleiss*). Nonetheless he maintained that it would join *The Creation* in assuring his lasting fame. For the publication he took the path of lesser resistance, selling the rights to Breitkopf & Härtel.

Other than masses, Haydn's only important liturgical work from this period is the *Te Deum* 'for the Empress' (HXXIIIc:2), probably composed in 1800 and apparently first given in September in Eisenstadt, perhaps in conjunction with the visit there of Lord Nelson (whence the nickname 'Nelson' Mass for the *Missa in angustiis*). A very different kind of vocal composition is represented by the 13 partsongs (HXXV), composed in 1796–9. A number of lieder and canons date from the same years; the latter were also 'private' works, the autographs of which Haydn framed and mounted on the walls of his house. A chapter in its own right is the hundreds of arrangements of British folksongs he sold to the publisher George Thomson in Edinburgh, not all from his own pen; as in so many other respects, Beethoven followed him in this lucrative commission.

In 1799 Haydn began to complain of physical and mental weakness. He wrote to Härtel in June:

> Every day the world compliments me on the fire of my recent works, but no one will believe the strain and effort it costs me to produce them. Some days my enfeebled memory and the unstrung state of my nerves crush me to the earth to such an extent that I fall prey to the worst sort of depression, and am quite incapable of finding even a single idea for many days thereafter; until at last Providence revives me, and I can again sit down at the pianoforte and begin to scratch away.

Indeed his productivity began to decline about this time, although his music continued to gain in 'fire' and cogency as long as he continued to compose. His last major completed work was the Mass in B♭ (H XXII:14) 'Harmoniemesse', given in September 1802; there followed only (perhaps) the *Hungarischer National Marsch* (H VIII:4; referred to in a letter of November) and the unfinished String Quartet op.103 of 1802–3, originally intended to go with op.77 to complete an opus of three (published 1806, dedicated to Count Moritz von Fries). He deeply regretted his loss of the stamina and concentration necessary for composition. On 6 December 1802 he thanked Pleyel for the receipt of the latter's edition of his complete string quartets, adding: 'I only wish that I could have back 10 years of my advanced age, so that I could provide you with something new of my composition – perhaps – despite everything – it can still happen'; a draft or alternative version (printed by Griesinger) of the letter to Härtel quoted above begins: 'It is almost as if with the decline in my mental powers, my desire and compulsion to work increase. O God! how much remains to be done in this glorious art, even by such a man as I have been!' One inevitably thinks of Beethoven, who would soon prove him right: their relationship was difficult around the years 1798–1803, owing to Beethoven's 'anxiety of influence' vis-à-vis Haydn, the crisis of his deafness, Haydn's increasing frailty and their mutual ambivalence regarding Beethoven's eventual assumption of the mantle of greatest living composer.

In the same letter Haydn also complained about the press of business (he was then arranging the self-publication of *The Creation*). Härtel added to that burden by initiating negotiations towards the publication of the so-called *Oeuvres complettes* (1800–6). This was not a complete edition but comprised saleable works for or with piano, analogous to an edition of Mozart already under way. In this connection Härtel sent Haydn a list of compositions attributed to him, asking him to pronounce on their authorship; Haydn (rather casually) did so. This was the first of several initiatives that led to Haydn's and Elssler's production in 1805 of the comprehensive 'Catalogue of all those compositions that I approximately recall having composed from my 18th to my 73rd year', or 'Haydn-Verzeichnis'.

In 1800 Haydn's wife died, leaving much of her modest estate to him. No longer interested in marriage, he signed a declaration in which he promised to marry Luigia Polzelli or no one, as well as to bequeath her 300 gulden a year (a sum he later reduced to 150). In May–June 1801 he drew up his will; the largest bequests were to his two brothers, but he also provided for many other relatives, former benefactors (and perhaps lovers), servants, religious and charitable institutions etc. From 1800 on he received a steady stream of distinguished visitors, honours and medals of which a gold medal from and honorary citizenship of the City of Vienna, in recognition of his charitable performances, meant the most to him. He also continued to talk shop with younger musicians; in addition to Pleyel and Beethoven, those who benefited included Anton Wranitzky, Neukomm, Reicha, Eybler, Weigl, Seyfried, Hummel, Diabelli, Kalkbrenner and Weber.

Haydn's last public musical function was on 26 December 1803, when he conducted the *Seven Last Words*. Thereafter he mainly stayed at home in Gumpendorf; after 1805 he wrote no letters in his own hand. His youngest brother Johann died in Eisenstadt in that year, and Michael (who in 1801–2 had declined an offer to succeed him as Esterházy Kapellmeister) followed in 1806. His last public appearance of any kind was on 27 March 1808, at a gala performance of *The Creation* in honour of his 76th birthday, in the Great Hall of the University. He signed his last will on 7 February 1809, altered primarily to reflect the many deaths that had taken place in the meantime. It disbursed about 24,000 gulden; his estate totalled about 55,000 gulden. During the spring he progressively weakened and had to be cared for by Elssler and other servants. His final decline was hastened by the French bombardment of Vienna on 11–12 May; his last known visitor was a French officer who paid his respects and sang an aria from *The Creation* (although the 'honour guard' ordered by Napoleon seems to be a legend). After playing the 'Emperor's Hymn' on 26 May 'with such expression and taste that our good Papa was astonished about it himself ... and was very pleased' (Elssler), he had to be assisted to bed; he did not rise again. He died at about 20 minutes before 1 a.m. on 31 May; owing to the war only a simple burial was possible, the following afternoon. On 15 June a solemn memorial service was held in the Schottenkirche, with a performance of Mozart's Requiem. His remains are now interred in the Bergkirche in Eisenstadt.

6. CHARACTER AND PERSONALITY

The traditional image of Haydn's personality has been that of 'Papa Haydn': pious, good-humoured, concerned for the welfare of others, proud of his students, regular in habits, conservative. Although not inaccurate, it is one-sided; it reflects the elderly and increasingly frail man his first biographers knew. Insight into the personality and behaviour of the vigorous and productive composer, performer, Kapellmeister, impresario, businessman, conqueror of London, husband and lover, whose career had already spanned 50 years when Griesinger met him in 1799, must be inferred from his correspondence (which is more revealing than is usually assumed) and from other sources.

Haydn's public life exemplified the Enlightenment ideal of the *honnête homme*: the man whose good character and worldly success enable and justify each other. His modesty and probity were everywhere acknowledged (he was occasionally entrusted with secret diplomatic communications). These traits were not only prerequisites to his success as Kapellmeister, entrepreneur and public figure, but also aided the favourable reception of his music. A more appropriate sense of 'Papa' would be that of 'patriarch', as in the resolution making him a life member of the Tonkünstler-Societät in 1797, 'by virtue of his extraordinary merit as the father and reformer of the noble art of music'. The many younger musicians who benefited from his teaching and advice seem to have regarded him as a father-figure, or kindly uncle; he was well-disposed towards them and helped them as much as he could, although there is little evidence that he thought of them as substitutes for the children he (apparently) never had.

When conditions permitted (i.e. in Vienna and London) Haydn enjoyed a rich emotional and intellectual life. In addition to his intimate relationships with Polzelli, Genzinger and Schroeter, he developed warm friendships with Mozart and Albrechtsberger; with Burney, Dr and Mrs Hunter and Therese Jansen; with an unnamed man to whom he wrote from Vienna in December 1792 (in English), playing the role of honorary godchild: 'I rejoice very much that my handsome and good Mother Susana has changed her state ... I wish from all my heart, that my Dear Mother may at my arrival next year present me a fine little Brother or Sister'; with 'my dear' Nancy Storace and 'my very dearest' Nanette Bayer (a 'great genius' of a pianist, employed by Count Apponyi) and many others. His observations in the London Notebooks reveal an active interest in every aspect of social life and culture, 'high' and 'low' alike. He was interested in literature, art and philosophy and gladly circulated and corresponded with intellectuals and freemasons, albeit without pretensions to being an intellectual himself.

Haydn's character was marked by a duality between earnestness and humour. F.S. Silverstolpe, who saw much of him during the composition of *The Creation*, reported:

> I discovered in Haydn as it were two physiognomies. One was penetrating and serious, when he talked about anything sublime, and the mere word 'sublime' was enough to excite his feelings to visible animation. In the next moment this air of exaltation was chased away as fast as lightning by his usual mood, and he became jovial with a force that was visible in his features and even passed into drollery. The latter was his usual physiognomy; the former had to be induced.

The many anecdotes about Haydn's youthful propensity to practical joking, however implausible individually, must collectively reflect some reality. Griesinger found that 'a guileless roguery, or what the British call *humour*, was one of Haydn's outstanding characteristics'. But he was also a devout Catholic: he inscribed most of his autographs 'In nomine Domini' at the head and 'Laus Deo' at the end, and composed major works in honour of the Virgin, including the *Stabat mater*, the *Salve regina* in G minor, and the *Missa Cellensis* and 'Grosse Orgelmesse'. His most important instrumental work of the 1780s was arguably the *Seven Last Words*. He identified personally with *The Creation* and the religious portions of *The Seasons* and came to think of the former in overtly moralistic terms, as he wrote in 1801:

> The [story of the] Creation has always been considered the sublimest and most awe-inspiring image for mankind. To accompany this great work with appropriate music could certainly have no other result than to heighten these sacred emotions in the listener's heart, and to make him highly receptive to the goodness and omnipotence of the Creator.

Haydn's personality was more complex than has usually been thought. His marriage was unhappy, and he was often lonely and at times melancholy. In May 1790 he wrote to Mme Genzinger: 'I beg Your Grace not to shy away from comforting me by your pleasant letters, for they cheer me up in my isolation, and are highly necessary for my heart, which is often very deeply hurt'. Nor was it only a question of his physical and social isolation at Eszterháza; from London he wrote to Polzelli of his 'melanconia' in much the same terms.

His modesty, genuine though it was, had distinct limits. He took pride in his works, notably including his vocal music; he wrote to Artaria in May 1781:

> My lieder, through their variety, naturalness, and beautiful and grateful melodies, will perhaps surpass all others ... Now something from Paris: Mr leGros ... wrote me all sorts of nice things about my *Stabat mater*, which was given ... to the greatest applause ... They were surprised that I was so extraordinarily successful in vocal music; but *I* wasn't surprised at

all, for they haven't heard anything yet. If they could only hear ... my most recent opera *La fedeltà premiata*! I assure [you] that nothing comparable has yet been heard in Paris, and perhaps not even in Vienna.

Haydn prized his status as an original; he bluntly rejected the notion that Sammartini might have been an influence on his early string quartets, adding (to Griesinger) that he acknowledged only C.P.E. Bach as a model. He was sensitive to criticism: he resented North Germans' rejection of his stylistic mixture (which seemed to them a breach of decorum), was jealous of Joseph II's patronage of inferior composers such as Leopold Hofmann (whose 'Gassenlieder' in particular he intended to surpass), and railed against those who pointed out technical flaws such as parallel fifths in his late music. He took pains to forestall potential criticism of the similar beginnings of the sonatas HXVI:36/ii and 39/i by printing a prefatory note asserting that he had done this deliberately, 'in order to show different methods of realization'.

After Mozart's death he willingly accepted the role of greatest living composer; in London he actively defended his 'rank' against Pleyel's challenge. Despite his grateful dependence on Swieten for librettos and patronage by the Associirten, behind Swieten's back he gibed that his symphonies were 'as stiff as the man himself' and ridiculed the libretto of *The Seasons*. His despair at no longer being able to compose after 1802 was doubtless fuelled in part by resentment at Beethoven's success in pushing forward into new domains of music – domains that he believed would have lain open to him if only his health had not failed.

Although Haydn often protested his devotion to the Esterházy princes, 'in whose service I wish to live and die' (1776), and praised Nicolaus for providing the conditions under which his art could develop, his attitude towards the court was never subservient and over time became increasingly ambivalent. He did not hesitate to assert his interests and those of his musicians against the court administration: these interventions were usually successful (including, for example, re-engaging Polzelli and her husband and keeping them on the payroll for ten years). The relationship between him and Nicolaus was not merely that of prince and employee: their playing baryton trios together (Haydn presumably on the viola) was by definition intimate music-making, and according to Framery he had to restore the prince from attacks of depression (Haydn himself described this in March 1790, admittedly during Nicolaus's bereavement). After 1780 he became increasingly independent of the court both compositionally and financially, and he hated having to abandon the artistic and social pleasures of Vienna for distant Eszterháza.

After Nicolaus's death he was *de facto* a free artist. In a letter of September 1791 to Mme Genzinger his ambivalence is palpable:

> This little bit of freedom, how sweet it tastes! I had a good prince, but at times I was forced to be dependent on base souls. I often sighed for release; now I have it in some measure ... Even though I am burdened with more work, the knowledge that I am not bound to service makes ample amends for all my toil. And yet, dear though this freedom is to me, I long to be in Prince Esterházy's service on my return, if only for the sake of my poor family. However, I doubt that this longing can be satisfied, in that my prince ... absolutely demands my immediate return, which however I cannot comply with, owing to a new contract I have entered into here.

Indeed, London won out. Yet in 1795 he was glad to become Kapellmeister again – although only on condition of minimal duties.

Haydn's realism was most obvious regarding money. Haydn was so self-interested as to shock both certain high-minded contemporaries (Joseph Martin Kraus, Friedrich Rochlitz) and many later authorities. Whereas until 1749 he presumably suffered nothing worse than ordinary schoolboy privations, during his early freelance years he lived in poverty, an experience he was determined never to repeat. He always attempted to maximize his income, whether by negotiating the right to sell his music outside the Esterházy court, driving hard bargains with publishers or selling his works three and four times over; he regularly engaged in 'sharp practice' and occasionally in outright fraud. When crossed in business relations, he reacted angrily. At times his protestations of straitened circumstances were mendacious (as when denying Polzelli's requests for money), or perhaps self-deceptive. Yet Haydn was generous. He supported his brother Johann for decades and bequeathed substantial sums to relatives, servants and those who had supported him in his youth, and took pride in the large sums generated for charity by performances of his oratorios.

Haydn's appearance is known from various descriptions and from many paintings and busts. He was not handsome; he was 'small in stature, but sturdily and strongly built. His forehead was broad and well modelled, his skin brownish, his eyes lively and fiery, his other features full and strongly marked, and his whole physiognomy and bearing bespoke prudence and a quiet gravity' (Griesinger); but he also had an overlarge nose, exacerbated by his long-term polyp, and was pock-marked (Dies). Of the many contemporary images only a few avoid idealizing their subject. From about 1768 there is a portrait by Grundmann, the

Esterházy court painter, showing the young and self-assertive Kapellmeister in uniform. More conventional in both facial features and the pose at the keyboard are the various images based on a lost painting of uncertain date by Guttenbrunn (Haydn's wife and Griesinger claimed that it was a good likeness) and the engraving by Mansfield (1781) published by Artaria. There is a good miniature from about 1788. From London we have formal portraits by Hoppner and Hardy, of which the former has the more personality; still more is conveyed in the drawing by George Dance (in two versions, which Haydn claimed was the best likeness of himself. Several sculptures survive from the last Vienna years, including two busts by Grassi (praised by Griesinger); there is also a deathmask, taken by Elssler.

7. STYLE AND METHOD

Haydn's style was understood in his own day as unique. He famously commented to Griesinger:

> My prince was satisfied with all my works; I received approval. As head of an orchestra I could try things out, observe what creates a [good] effect and what weakens it, and thus revise, make additions or cuts, take risks. I was cut off from the world, nobody in my vicinity could upset my self-confidence or annoy me, and so I had no choice but to become original.

By 'original' he seems to have meant that he belonged to no school and acknowledged few if any models. However, in late 18th-century aesthetics originality also implied genius, a link emphasized among others by Kant.

In many ways Haydn's style can be interpreted in terms of the duality in his personality between earnestness and humour. He said as much when referring to his method of composition: 'I sat down [at the keyboard] and began to fantasize, according to whether my mood was sad or happy, serious or trifling'. Of course, in his music these qualities are not unmediated binary opposites but poles of a continuum. Admittedly, since about 1800 wit has been the better understood pole. Johann Karl Friedrich Triest wrote (1801) of his 'unmistakable manner': 'what the English call "humour", for which the German *Laune* does not quite provide an exact equivalent'. Haydn's 'unique' or 'inimitable' *Laune* was a frequent motif in contemporary criticism. Most of the familiar nick-names for his works respond to features that listeners have taken as humorous; e.g., the 'Surprise' Symphony or the 'Joke' Quartet, op.33 no.2. In other cases the wit is on a higher plane, e.g. the 'ticking' accompaniment in the slow movement of the 'Clock' Symphony, no.101. The crucial point, however, is that Haydn's popular style is not a simple projection of his personality, but his compositional 'persona' or 'musical personality', deliberately assumed for complex artistic purposes. Indeed 'wit' signifies intelligence as well as humour: his inexhaustible rhythmic and motivic inventiveness, the conversational air of many quartet move-ments, his formal ambiguity and caprice, his brilliant and at times disquieting play with beginnings that are endings and the reverse (the 'Joke' Quartet ending has stimulated half a dozen learned exegeses). Often Haydn's wit shades into irony, as was recognized by his contempo-raries: 'Haydn might perhaps be compared, in respect to the fruitfulness of his imagination, with our Jean Paul [Richter] (omitting, obviously, his chaotic design; transparent representation (*lucidus ordo*) is not the least of Haydn's virtues); or, in respect to his humour, his original wit (*vis comica*) with Lor. Sterne' (Triest). In fact, his irony goes beyond wit: a passage may be deceptive in character or function (the D major inter-lude in the first movement of the 'Farewell' Symphony sounds like a minuet out of context, but it is not a minuet and plays a crucial tonal and psychological role), or a movement may systematically subvert listeners' expectations until (or even past) the end (the finale of the Quartet op.54 no.2). Like Beethoven, Haydn often seems to problema-tize music rather than merely to compose it (the tonal ambiguity at the beginning of op.33 no.1).

Earnestness and depth of feeling are equally important to Haydn's art. These qualities were less appreciated in the 19th and early 20th centuries, owing in part to the absence of his vocal music and much of his earlier instrumental music from the standard repertory, in part to a lack of sympathy for his extra-musical and ethical concerns during the age of absolute music. But Griesinger reported: 'Haydn said that instead of so many quartets, sonatas and symphonies he should have composed more vocal music, for he could have become one of the leading opera composers'. Until about 1800 vocal music was as responsible for his reputation as instrumental; Gerber wrote in 1790: 'around the year 1780 he attained the highest level of excellence and fame through his church and theatre works'. Like all 18th-century composers, Haydn believed that the primary purpose of a composition was to move the listener, and that the chief basis of this effect was song. He was an excellent tenor in the chamber (if not the theatre). He insisted to Griesinger that a prerequisite for good music was 'fluent melody', and he 'criticized the fact that now so many musicians compose who have never learnt how to sing. "Singing must almost be reckoned one of the lost arts; instead of song, people allow the instruments to dominate"'.

This emphasis on feeling also applies to instrumental music – even sprightly allegros and minuets – and throughout Haydn's career. Much of his early music is earnest, at times even harsh; see the Keyboard Trio HXV:f1, the String Trio HV:3, the slow movement of the String Quartet op.2 no.4, Symphony no.22 and much else, to say nothing of vocal works such as the *Stabat mater* and the *Salve regina* in G minor. Many of his keyboard works are affective in an intimate way: he wrote to Mme Genzinger regarding the Adagio of Sonata no.49: 'It means a great deal, which I will analyse for you when I have the chance'. His orchestral music 'signified' as well: the slow introductions to the London symphonies are implicit invocations of the sublime, and this topic became overt in the Chaos-Light sequence in *The Creation* and elsewhere in his late sacred vocal music. Many works that were later taken as humorous he did not intend as such, for example the 'Farewell' Symphony. Similarly, even at his wittiest or most programmatic he never abandons tonal and formal coherence.

The duality between earnestness and wit is analogous to the 18th-century distinctions between connoisseurs ('Kenner') and amateurs ('Liebhaber'), and between traditional or learned and modern or *galant* style. These dualities characterize many of Haydn's works, groups of works and even entire periods. In his pre-Esterházy instrumental music, genre was a primary determinant of style: modest, unpretentious divertimentos, quartets and keyboard concertinos etc. stand seemingly opposed to larger-scale symphonies, string trios and keyboard trios. The three op.20 quartets with fugal finales project, in order of composition, severe tradition (no.5), the *galant* (no.6) and a studied mixture of both (no.2); yet these monuments to high art originated precisely in the middle of his baryton-trio decade. In the late 1770s most of his symphonies were unambiguously intended as entertainment, but no.70 is selfconsciously learned. In 1785–90 he composed some 45 weighty symphonies, quartets and piano works, but also lyre concertos and noturnos, flute trios and other light works. Of course, the distinction between 'art' and 'entertainment' cannot be simplistically correlated with differences in artistic quality. Haydn's early string quartets are arguably his most polished pre-Esterházy works; the baryton trios and lyre notturnos are finely wrought compositions, as rewarding in their way as the raw expressionism of the 'Sturm und Drang'. These stylistic dualities are found even in his late sacred vocal music and long hindered its appreciation. His quotation of the *buffa*-like contredanse from no.32 of *The Creation* in the 'Schöpfungsmesse' so offended the empress that she insisted that he alter it in performances at the Habsburg court, many of her high-minded contemporaries took offence at the 'Tändeleien' (trifling) and dance-like triple metres in his late masses, and as recently as the 1970s noted authorities still wrote of the 'triviality' of the Kyrie of the *Missa in tempore belli*. Now, however, their stylistic heterodoxy seems as gloriously uplifting as that of *Die Zauberflöte*.

Haydn usually juxtaposes or contrasts stylistic dualities rather than synthesize them. Perhaps he approaches synthesis most closely when an ostensibly artless or humorous theme later changes in character (e.g. Symphony no.103, minuet) or is subjected to elaborate contrapuntal development; the latter is especially characteristic of finales (e.g. Symphony no.99; Beethoven twice copied out the development section). In general, Haydn's art is based on the traditional principle of variety within unity. 'Once I had seized upon an idea', he said to Griesinger, 'my whole endeavour was to develop and sustain it in keeping with the rules of the art'. A Haydn movement works out a single basic idea; the 'second theme' of his sonata forms is often a variant of the opening theme. Often this part of the exposition forswears thematic statements altogether, in favour of unstable developmental passages (his 'expansion section'); stability is restored only in the position of the usual closing theme. To be sure, that working out usually entails many contrasting treatments and effects (Haydn: 'light and shade', i.e. chiaroscuro): the second theme usually differs in treatment, and the recapitulation brings fresh developments; in his double-variation slow movements the alternating major and minor themes are usually variants of each other. Thus both novelty and continuity are maintained from beginning to end.

In one respect, however, Haydn deliberately courted a union of opposites: his 'popular' style that simultaneously addressed the connoisseur. 'If one wanted to describe the character of Haydn's compositions in just two words, they would be ... *artful popularity* or *popular* (easily comprehensible, effective) *artfulness*' (Triest). No other composer – not even C.P.E. Bach or Mozart – had Haydn's gift of writing ostensibly simple or folklike tunes of wide appeal, and broadly humorous sallies, that concealed (or developed into) the highest art. Indeed these aspects of his style intensified in his London and late Vienna years, along with the complexity of his music and its fascination for connoisseurs. One of the best early comments on Haydn's music was Gerber's: he 'possessed the great art of *appearing* familiar in his themes' (emphasis added): that is, their popular character is neither merely given nor a direct reflection of his personality, but the result of calculated artistic shaping. This becomes obvious when he employs folk tunes, as in the Andante of Symphony no.103 and the finale of no.104: the piquant raised fourth-degree of the one, the horn pedal of the other, are not quoted, but adapted to the character of a grand symphony. Haydn's 'pretension ... to a simplicity that appears to come from Nature itself is no mask but the true claim of a style whose command over the whole range of technique is so great that it can ingenuously afford to disdain the outward appearance of high art' (Rosen, I1971).

Many aspects of Haydn's music can be appreciated only by ignoring the concept of 'Classical style'. These include lean orchestration (Haydn: 'no superfluous ornaments, nothing overdone, no deafening accompaniments'), in which the planes of sound do not compactly blend but remain distinct, nervous bass lines, constant motivic-thematic development and a rhythmic vitality and unpredictability that can become almost manic, as in the finales of many late string quartets and piano trios. Many Haydn movements are progressive in form, continually developing (e.g. the first movements of Symphonies nos.92 and 103); on a still larger scale, many works exhibit tendencies towards through-composition or 'cyclic' organization; a few are as tightly integrated as any work of Beethoven (e.g. the 'Farewell' Symphony and no.46; the String Quartets op.20 no.2, op.54 no.2 and op.74 no.3; Piano Sonata no.30).

Haydn was also a master of rhetoric. This is a matter not only of musical 'topoi' and rhetorical 'figures' but also of contrasts in register, gestures, implications of genre and the rhythms of destabilization and recovery, especially as these play out over the course of an entire movement. Referential associations are common in his instrumental music, especially symphonies (nos. 6–8, 22, 26, 30–31, 44–5, 49, 60, 64, 73, 100); they invoke serious human and cultural issues, including religious belief, war, pastoral, the times of day, longing for home, ethnic identity and the hunt. Haydn told Griesinger and Dies that he 'often portrayed moral characters in his symphonies' and that one early Adagio presented 'a dialogue between God and a foolish sinner' (unidentified; perhaps from no.7, 22 or 26). In his vocal music Haydn (like Handel) was a brilliant and enthusiastic word-painter. This trait is but one aspect of his musical imagery in general: in addition to rhetorical figures and 'topoi' it comprises key associations (e.g. E♭ with the hereafter), semantic associations (e.g. the flute with the pastoral) and musical conceptualizations (e.g. long notes on 'E-wigkeit' in *The Creation* or 'ae-*ter*-num' in the late *Te Deum*).

Like all 18th-century composers, Haydn composed for his audiences (which term includes his performers). He calculated Piano Sonata no.49 expressly for Mme Genzinger; in his piano works of 1794–6 he systematically differentiated between a difficult, extroverted style for Therese Jansen and a less demanding, intimate one for Rebecca Schroeter. Regarding the piano trio HXV:13 he wrote to Artaria: 'I send you herewith the third trio, which I have rewritten with variations, to suit your taste' – i.e. Artaria's estimate of the taste of Haydn's market. When he went to London, his music for public performance became grander and more brilliant. He disliked having to compose without knowing his audience, as he wrote regarding *Applausus*: 'If I have perhaps not divined the taste of [the musicians], I am not to be blamed for this; neither the persons nor the place are known to me, and the fact that they were concealed from me truly made my work distasteful'.

'I was never a hasty writer, and always composed with deliberation and diligence', Haydn told Griesinger. His method encompassed three stages: 'phantasieren' at the keyboard in order to find a viable idea, 'komponieren' (working out the musical substance, both at the keyboard and by means of shorthand drafts, usually on one or two staves) and 'setzen' (writing the full score). Sketching was a regular procedure: although drafts survive for only a modest proportion of his music, they comprise works in all genres and all types of musical context (including recitatives). A draft for the finale of Symphony no.99 confirms Griesinger's description of his use of numbered cross-references to organize a series of passages originally written down in a different order. His surviving autographs by and large are fair copies, which exhibit few corrections and alterations.

8. THE WORKS

SACRED VOCAL MUSIC

Vocal music constitutes fully half of Haydn's output. Both his first and last completed compositions were mass settings, and he cultivated sacred vocal music extensively throughout his career except during the later 1780s, when elaborate church music was inhibited by the Josephinian reforms, and the first half of the 1790s in London.

The *Missa brevis* in F (HXXII:1) is apparently his earliest surviving composition; on rediscovering it in old age he pronounced himself pleased by 'the melody and a certain youthful fire' (Dies), which are enhanced by resourceful contrasts between the two solo sopranos and the chorus. The remaining masses fall into two groups of six each: nos.2, 4–8 (1766–82; no.3 is probably spurious) and nos.9–14 (1796–1802); except where noted they are of medium length (30 to 40 minutes). The former are notably heterogeneous. The huge and impressive *Missa Cellensis* in C (begun 1766) is of the *solenne* type (often miscalled 'cantata mass'); each of the five main sections is subdivided into numerous complete and independent movements. These include choruses both festive and ominous, elaborate arias, ariosos, ensembles and four massive concluding fugues. The Kyrie and certain arias are traditional in style, while the remainder is distinctly modern; the fugues are powerfully expressive despite their contrapuntal fireworks, especially the overwhelming 'Et vitam venturi', which functions not merely as a concluding highpoint but as the through-composed goal of the entire Credo. The *Missa sunt bona mixta malis* (1768) survives only in an autograph fragment transmitting the Kyrie and the first part of the Gloria; it is not known whether Haydn completed the work, and the import of 'mixed good and bad' (from a classical proverb) remains obscure. It is set for chorus and organ continuo in 'stile antico'; strict fugal expositions alternate with free counterpoint and occasional homophonic passages. The Grosse Orgelmesse in E♭ (c1768–9) is more personal in tone: the dark english horns contrast with exuberant treble obbligato organ parts in the Kyrie, Benedictus and 'Dona nobis pacem'. The *Missa Sancti Nicolai* (1772) is often described as 'pastoral', owing to its key of G major and the lilting 6/4 rhythm of the Kyrie (which returns for the 'Dona nobis pacem'), although the Crucifixus and Agnus Dei are serious indeed. In the mid-1770s followed the 'Little Organ Mass' in B♭, a quiet, almost pietistically fervent *Missa brevis*. The 'Mariazellermesse' in C (1782) resembles the *Missa Cellensis* in key, scoring and purpose, although it is more compact and more closely allied with sonata style.

Not withstanding their semi-private function for the Esterházy court, Haydn's six late masses are consummate masterworks that exhibit no trace of provinciality or the 'occasional'. He exploits the complementary functions of soloists and chorus with inexhaustible freedom and telling effect; owing to his London experience the orchestra plays a newly prominent role. Four are in B♭, perhaps because bb'' was Haydn's usual highest pitch for choral sopranos (he employed the same key for the final choruses of Parts 2–3 of *The Creation* and Part 1 of *The Seasons*). The other two are the only ones for which he provided descriptive titles: the *Missa in tempore belli* ('Mass in Time of War', 1796) in C features the bright, trumpet-dominated sound typical of masses in this key; the *Missa in angustiis* ('Mass in [times of] distress', later nicknamed 'Nelson' Mass, 1798) in D minor and major is scored for a dark orchestra comprising only trumpets and timpani, strings and organ. Both invoke the travails of the Napoleonic wars. The Agnus Dei of the former includes threatening timpani motifs and harsh trumpet fanfares, while the Benedictus of the latter culminates in another harsh fanfare passage 'out of context'; both influenced the Agnus Dei in Beethoven's *Missa solemnis*. On the other hand, except for the sombre Kyrie and Benedictus of the 'Nelson' Mass, both are otherwise firmly optimistic; the ending of the latter is downright jaunty.

Although Haydn's late masses indubitably reflect the experience of the London symphonies, their symphonic character has been exaggerated. Even in the Kyrie, which usually consists of a slow introduction and a fast main movement, the latter freely combines fugato and sonata style in a distinctly unsymphonic way. The Gloria and Credo are divided into several movements, fast–slow–fast with the slow middle movement(s) in contrasting keys and featuring the soloists (e.g. the 'Qui tollis' of the *Missa in tempore belli*, a bass aria with solo cello in A major; or the 'Et incarnatus' of the 'Heiligmesse', based on Haydn's canon *Gott im Herzen*); they usually conclude with a fugue on a brief subject, which often enters *attacca* and always leads to a homophonic coda. The Sanctus often adopts the 'majesty' topic, admixed with mysterious passages; it leads directly into the brief 'Pleni sunt coeli – Osanna', which may or may not return following the Benedictus. The latter is a long movement and an emotional highpoint; it usually features the soloists and is in, or based on, sonata form. The Agnus Dei opens with an initial slow section, either threatening in the minor or serenely confident in a remote major key; it leads to a half-cadence and thence to the fast 'Dona nobis pacem', usually a free combination of fugato and homophony, leading (again) to a homophonic wind-up.

The other liturgical works date primarily from the first half of Haydn's career; their original destinations and purposes are almost entirely unknown. According to liturgical function they comprise offertories (H XXIIIa), Marian antiphons (H XXIIIb), hymns (H XXIIIc) and pastorellas (H XXIIId; Haydn called them 'cantilenas'). They vary widely in style and scale, from the massive, dark, traditional *Stabat mater* (H XXbis, 1767) to the tender devotion of the *Lauda Sion* hymn-complexes; from the festive jubilation of the choral *Te Deum* settings with trumpets and drums in C to the stylized folk idiom of the pastorellas for solo voices and strings. Even subgenres exhibit marked contrasts: the *Lauda Sion* hymns from the 1750s (HXXIIIc:5) are all in C, Vivace 3/4, while those from the later 1760s (HXXIIIc:4) are in a tonally interesting set of four different keys and alternate Andante 3/4 with Largo alla breve. Similarly, the *Salve regina* in E (HXXIIIb:1, 1756) features ornate italianate writing for the solo soprano, whereas that in G minor (HXXIIIb:2, 1771) is expressively brooding, with no trace of vocal ornamentation. Of the three late works, the offertory *Non nobis, Domine* in D minor (HXXIIIa:1, ?1780s) is an *a cappella* work reminiscent of the *Missa sunt bona mixta malis*, while the six English Psalms of 1794 (HXXIII, Nachtrag), Haydn's only Protestant church music, adumbrate the elevated but plain style of 'The Heavens are Telling' of *The Creation*. The late *Te Deum* 'for the Empress' (HXXIIIc:2, ?1800), for chorus and very large orchestra, is an *ABA* construction of great power and terseness; it whirls through the very long text in little more than eight minutes, while still finding time for a double fugue and an immense climax at the end.

Haydn's oratorios comprise *Il ritorno di Tobia*, his revision of Friebert's arrangement of the *Seven Last Words*, *The Creation* and *The Seasons*. The libretto of *Tobia* (by a brother of Boccherini) narrates the story of the blind Tobit from the Apocrypha; Haydn fashioned a magnificent late example of Austrian-Italian vocal music, comprising chiefly long bravura arias, along with three choruses; most of the recitatives are *accompagnati* of emotional intensity. In 1784 he revived the oratorio, shortening many of the arias, adding two magnificent new choruses and supplementing the instrumentation. The *Seven Last Words*, a success during Haydn's lifetime and beyond, is less popular today, in part because it is not a full-length work, in part owing to the succession of eight consecutive adagios which, paradoxically, seem more monotonous than in the orchestral version. Its most striking movement is the bleak, newly-composed introduction to the second part, scored for wind alone and set in A minor, a key Haydn hardly ever used.

The Creation is Haydn's most loved work today, as it was in his lifetime. Part 1 treats the First to Fourth Days (the creation of light, land and sea, plant life, heavenly bodies), Part 2 the Fifth and Sixth (animals, birds, fish, man and woman); each Day comprises recitative on prose from Genesis, a commentary set as an aria or ensemble, another recitative and a choral hymn of praise. Part 3 abandons the Bible; it amounts to a cantata devoted to Adam and Eve and to further praise of Heaven. The optimistic tone is enhanced by the increasing brilliance and complexity of the choruses as the work proceeds; they reflect Haydn's experience of Handel in England. Also reminiscent of Handel (not that Haydn needed the stimulus) are the many word- and scene-paintings, of which the most striking include the emergence of the oceans and mountains ('Rolling in foaming billows'), the sunrise and moonrise, the birds of 'On mighty pens' and the teeming low strings of 'Be fruitful all'; though often taken as humorous, these conceits are essential to the Enlightenment optimism of the work. The famous 'Representation' or 'Idea of Chaos' (*Vorstellung* implies both meanings) is not literally chaotic but paradoxical: beginning in C minor mystery, it initiates a larger process which points beyond itself, and acquires meaning only with the choral climax on 'And there was light!' in C major. The remainder of Part 1 takes place, as it were, during the reverberation of this event; its triumphant concluding chorus 'The Heavens are Telling' is again in C. By contrast, the final sections dealing with ourselves shift to the 'human' key of Bb and its sub-dominant Eb. Although Part 3 opens in a radiant, astonishingly remote E major for the Garden of Eden, it soon reverts to F and C for the gigantic 'Lobgesang' and, via Eb for Adam's and Eve's lovemaking in earthy Singspiel style, to Bb for the final choral fugue.

The Seasons libretto presents scenes of nature and country life; the narrator-function is personified as the moralizing peasants Simon, Jane and Lucas. The scenic aspects stimulated Haydn to his best efforts: the storms of late winter, the farmer sowing his seed to the tune of the Andante of the 'Surprise' Symphony, a sunrise that outdoes that in *The Creation*, the thick C minor fogs of early winter, and the multi-movement depiction of summer heat, first languid, then oppressive, finally exploding in Haydn's greatest storm. Among the genre scenes those for the chorus are unsurpassed, notably at the end of Autumn: first the hunt, from sighting to chase to kill to celebration (the horns quote numerous actual hunting calls and join the trombones and strings in double grace-notes for the baying of the hounds), in progressive tonality from D to Eb; then the drinking chorus in C, with increasingly uncertain harmonizations of a prominent high note for the raising of glasses, a dance in 6/8 leading to an inebriated fugue and, finally, a breathless wind-up that may have inspired the end of Verdi's *Falstaff*. Other important choruses are pastoral ('Komm, holder Lenz') and religious: 'Ewiger, mächtiger, gütiger Gott' at the end of Spring, Haydn's most massive chorus (itself run on from the preceding trio, the two movements as a whole in 'progressive tonality'); and the concluding 'Dann bricht der grosse Morgen an', in which we enter Heaven in a blaze of C major glory, resolving the C minor of the beginning of Winter. Notwithstanding its less exalted subject, *The Seasons* is compositionally more virtuoso than *The Creation* and offers greater variety of tone: Haydn's pastoral is one of the final glories of a tradition that is more than 'high' enough.

In Haydn's sacred vocal music the aesthetics of through-composition is a matter not only of cyclic integration, but of doctrine and devotion. Many of these works are organized around the conceptual image of salvation, at once personal and communal, achieved at or near the end: a musical realization of the desire for a state of grace. This is especially clear in a relatively brief work such as the *Salve regina* in G minor, where the astonishing vocal entry on an augmented sixth chord is not really resolved until the end, when Haydn 'hears' the supplicants' prayer by turning to the major. Particularly in his late sacred music such concepts are wedded to the sublime: not only in the Creation of Light, which expresses that which is otherwise unthinkable – the origins of the universe and of history – but also in the choruses that conclude each part of *The Creation*, 'Spring' and 'Winter' in *The Seasons* and many movements of the late masses.

SECULAR VOCAL MUSIC

Haydn's stage works comprise 13 Italian operas, four Italian comedies (with spoken dialogue rather than recitative), five or six German Singspiele and incidental music for plays, of which only Symphony no.60, 'Il distratto', survives; almost all were composed for the Esterházy court. Those predating 1766 are lost, except for fragments of the *festa teatrale Acide* (1762, revised 1773) and of the *commedia Marchese (La Marchesa Nespola)* (1762–3). His three operas from the late 1760s become increasingly long and complex. The two-act intermezzo *La canterina* (1766) has wonderful comic scenes centring on the jealous singing teacher Don Pelagio and his charge Gasparina (who 'overreacts' to being thrown out of his house with a distraught aria in C minor); each act ends with a quartet. *Lo speziale* (1768), in three acts, is called a *dramma giocoso* and is based on a libretto of this type by Goldoni, but the Esterházy version eliminates the two *parti serie*. It has many new effects, including a 'Turkish' aria with 'exotic' key-relations and rhythms and a graphic portrayal of the effects of the apothecary's remedies for constipation. The concluding trio and quartet of the first and second acts, respectively, include real dramatic action. The three-act *Le pescatrici* (1769–70), also based on Goldoni, is a true *dramma giocoso* including the 'serious' Prince Lindoro and Eurilda, an heiress to a principality who has been raised as a simple fisherwoman; their music is in 'high' style, and Eurilda (in distinction to the eponymous fisherwomen) takes no part in the comic ensembles. It has more ensembles, in proportion to its total length, than any Haydn opera, although the majority are 'choruses' in primarily homophonic style. Among the latter is the Act 3 'Soavi zeffiri', whose E major tonality and depiction of sea breezes resemble Mozart's 'Placido è il mar' in *Idomeneo* and 'Soave sia il vento' in *Così fan tutte*.

After a pause, in 1773 Haydn composed *L'infedeltà delusa*, a 'burletta per musica' in two acts based on a libretto by the 'reforming' librettist Marco Coltellini. For the last time there are no serious characters; the opera portrays an idealized peasant life (with much lampooning of the nobility) and the characters are concerned only to set their mismatched affections aright. From the same year comes the German *Philemon und Baucis*, originally a marionette opera but surviving only in its version for the stage. The moralizing plot is based on the old theme of the god or king who is spiritually renewed by the incorruptible virtue of simple peasants; musi-cally it is similar to *L'infedeltà*, with the addition of impressive D minor music in the overture and thunderstorm chorus preceding Jupiter's arrival.

Most of Haydn's remaining operas for Eszterháza are in three acts and are *drammi giocosi* or other subgenres that mix comic and serious characters. In 1775 he composed *L'incontro improvviso* on a libretto adapted from Gluck's *La rencontre imprévue*; it is a harem-rescue plot set in the orient, as in Mozart's *Die Entführung aus dem Serail*, although many incidents lack sufficient motivation. The heroine Rezia and her rescuer Prince Ali are the serious characters, while lower-class characters provide broad 'exotic' humour. In a subplot Rezia uses her confidantes Balkis and Dardane to test Ali's fidelity (the gender-reversal is noteworthy); their Act 1 trio in the harem, with three sopranos sharing chromatic lines full of suspensions, is an invocation of timeless pleasure. In 1777 followed *Il mondo della luna*, based on Goldoni's popular libretto; the hero Ecclitico dupes the elderly Buonafede into supposing he has travelled to the moon (staged as an exotic, luxurious kingdom) and eventually into assenting to Ecclitico's marriage to his daughter Clarice (and two other marriages for good measure). The keys C and E♭ symbolize Earth and Moon respectively, the representation of the journey in the Act 1 finale being particularly magical, as is the Act 3 duet for the two principals. *La vera costanza* (1778–9, revised 1785), on a libretto by Francesco Puttini previously set by Anfossi, is Haydn's fullest exploration of the 'sentimental' subgenre of *opera buffa*. Rosina, secretly married to the half-mad Count Errico, lives incognito in a fishing-village. Eventually the Count and many other characters discover her, leading to repeated painful tests of her virtue and fortitude; in despair she flees to the country, where the final reconciliation takes place. The music is glorious and the characterizations surprisingly credible, with Rosina reaching heights of genuine emotion. The finales to Acts 1 and 2 are now (and largely remain) as long and complex as those in Mozart's operas.

A change of pace is represented by *L'isola disabitata* (1779), a relatively brief *azione teatrale* on a libretto by Metastasio, with all the recitatives orchestrally accompanied, and relatively brief, primarily lyrical arias without much coloratura. Next came *La fedeltà premiata* (1780), a *dramma pastorale giocoso* by G.B. Lorenzi, previously set by Cimarosa as *L'infedeltà fedele*; given the contrived plot-spring of the annual sacrifice of two lovers to appease an offended sea monster, the action and motivations are plausible. The number of arias in serious style is relatively high, with a climax in Celia's great scena in Act 2; the finale in Act 1 is Haydn's longest (822 bars). *Orlando paladino* (1782) is a *dramma eroicomico* with a libretto by Nunziato Porta based on Badini. Its subject is Orlando's madness (deriving ultimately from Ariosto's *Orlando furioso*), which Haydn portrays in remarkable scenes of mixed accompanied recitative and aria; the long scenes for Angelica and the feckless Medoro are musical highlights as well. *Armida* (1783) is a *dramma eroico* based on the Armida-Rinaldo action from Tasso's *Gerusalemme liberata*. It is primarily *seria* in style, with long stretches of action set in freely alternating accompanied recitatives and set-pieces; the long magic forest scene of Act 3 is particularly successful. *L'anima del filosofo, ossia Orfeo ed Euridice* (1791), an *opera seria* in four acts composed in London to a libretto by Badini, was not produced and possibly not completed. Notwithstanding numerous bravura arias, its style resembles that of Haydn's late instrumental works more closely than do his earlier operas; in Act 2 the extended scenes of Eurydice's death and Orpheus's discovery of her body are deeply affecting. It also includes numerous choruses, males for the Furies and females for the Bacchae, the latter ending the work (if that was the intention) in D minor.

During his career Haydn's operatic palette expanded both generically, from straight *buffa* or *seria* to various mixed types (reflecting the repertory as a whole), and compositionally, with longer individual numbers, interpenetration of accompanied recitative and set piece, and increased size and scope of the finale (except in *seria*). Notwithstanding his own high opinion of his operas, they were largely forgotten until the second half of the 20th century, when editions and recordings as well as stagings made them widely available. Their recent reception has been mixed. The music is beyond praise: the brilliance of Haydn's tonal and formal construction and his rhythmic verve go without saying; masterly too are his vivid characterization in arias, expressive strength in accompanied recitatives and fascinating orchestral effects; he often composes 'against the grain' of genre or the libretto to dramatic purpose. For these reasons (as well as their ready availability), they have attracted much analytical and critical attention. On the other hand, although the librettos represent major types and their thematic orientation is often strong, they often exhibit weaknesses of plausibility, motivation or dramaturgy; even Haydn's music cannot always overcome these faults, nor did he always exploit the dramatic implications of his librettos. For example, when deceptions are revealed in the Act 2 finales of *Lo speziale* and *Il mondo della luna*, the musical character does not change until later, when the people deceived (Sempronio, Buonafede) give vent to outbursts of rage; and Eurydice's second death remains anticlimactic (although here the libretto is also at fault). However, negative criticism has also been coloured by insufficient understanding of generic norms of the period 1760–80 (such as the dominance of aria over ensemble and 'seamless' action, and the relative brevity of the third act), and by inappropriate comparisons with Gluck and late Mozart instead of Gassmann, Anfossi or Cimarosa. In appropriate stagings with good singers, Haydn's operas are effective and moving in the theatre.

The festive Italian cantatas honouring Prince Nicolaus (H XXIVa:2–4, *c*1762–7) begin with a long orchestral ritornello leading to an accompanied recitative announcing the cause for celebration, followed by arias and duets and concluding with a chorus. The very long solo numbers are unusually virtuoso and richly orchestrated (in an aria from 'Qual dubbio ormai', no.4, Haydn wrote himself an elaborate obbligato harpsichord part). The celebratory cantata *Applausus* (H XXIVa:6, 1768) on an allegorical Latin text is stylistically similar, although it is longer and musically more concentrated, and as appropriate to its elevated text has been said to adumbrate the sublime. An important late chorus is *The Storm* (H XXIVa:8,1792); as in so many works of this type, minor-mode fury is followed by 'calm' in the major.

Three late solo cantatas for soprano are of great significance. *Miseri noi* (H XXIVa:7, by 1786) was composed for an unknown occasion and singer (possibly Nancy Storace); the middle section, a Largo in G minor, is particularly impressive. *Arianna a Naxos* (H XXVIb:2, ?1789) was perhaps composed for Bianca Sacchetti in Venice; in the passionate recitatives the piano presents the lion's share of the musical material, while the voice declaims the text dramatically. Ariadne's mixed hope and despair are vividly portrayed; in her final aria a long, slow, formal paragraph in F major leads to a wild rage aria in F minor, of which the final chord, for piano alone, is astonishingly F major. *Berenice, che fai* (H XXIVa:10,1795), on a text from Metastasio's *Antigono*, is public music for a virtuoso and hence more difficult and extroverted. The recitatives feature what is arguably Haydn's most extreme use of remote and enharmonic modulations; further the two arias are in 'opposed' keys (E major and F minor), while the orchestration is as brilliant as that of the last London symphonies.

Haydn's 47 songs (H XXVIa) comprise 24 German lieder (nos.1–24, 1781–4), 14 English songs (nos.25–36, 41–2, 1794–5, of which nos.25–36 were published as 'Canzonettas') and miscellaneous German lieder. The lieder of 1781–4 stand in close chronological and stylistic proximity to op.33. Although they have seemed simple to many commentators – they are relatively short and strophic, with the piano right-hand largely doubling the voice – they are varied in mood and exhibit subtle rhythmic and formal construction, often brilliantly realizing implications of the text; the 1784 set includes more deeply felt items. The English canzonettas contain many striking effects and are in many cases through-composed; see the remarkable off-tonic vocal entry of *She never told her love*, with its climax on 'smiling with grief', or the controlled passion of *O Tuneful Voice*: the poem invokes Mrs Hunter's sorrow at Haydn's departure, the music his farewell to her and to England. A special case is the 'Emperor's Hymn', with its fusion of elevated hymn and 'folk' styles. The 13 partsongs (*mehrstimmige Gesänge*; H XXVc, 1796–9) with keyboard accompaniment adumbrate the characteristic 19th-century Viennese genre of social music for vocal ensemble. Haydn said of them that they were composed '*con amore* in happy times and without commission' (Griesinger); as far as we know they (and his canons) are his only works of which this is true. They are among his wittiest, most beautiful and most touching creations, with an inimitable air of casual sophistication and a brilliant combination of comic and serious topics and styles; their fusion of easy intelligibility and wit with the highest art and their ravishing part-writing almost suggest string quartets for voices.

ORCHESTRAL MUSIC

Although Haydn's sobriquet 'father of the symphony' is not literally true, in a deeper sense it is apt: there is no other genre in Western music for which the output of a single composer is at once so vast in extent (106 works: HI:1–104, 107–8), so historically important and of such high artistic quality. His pre-Esterházy symphonies (most composed for Count Morzin) comprise nos.1, 37 and 18 (the earliest); 2, 4–5, 10–11, 17, 19–20, 25, 27, 32, 107; and possibly 3 and 15. All are scored for two oboes, two horns and strings; the majority are in three movements, fast–slow–fast. The distinction between a relatively weighty first movement and a faster finale is already present; the interior movement for strings alone is only moderately slow (Andante) and 'light' in style. Only nos.3 and 20 exhibit the later standard four-movement pattern; in nos.32 and 37 the minuet precedes the slow movement (found also in nos.108, 44, 68). In nos.5 in A and 11 in E♭ (the only ones in keys this distant from C), the slow movement comes first and is a weighty Adagio, producing the sequence slow–fast–minuet–fast with all four movements in the tonic (found also in nos.21–2, 34, 49). These early symphonies combine Italian and Austrian, light and serious, traditional and up-to-date features. Notwithstanding their limited outward dimensions, they are masterful; many exhibit considerable thematic integration (no.15) or manipulate generic norms to artistic effect (the opening movements of nos.15 and 25 are unusual in form, in ways that relate to the character and ordering of the succeeding movements); in no.3 the finale combines fugue and sonata form.

Haydn's years as Esterházy vice-Kapellmeister (1761–5) were his most productive as a symphony composer, with about 25 works (nos.6–9, 12–16, 21–4, 28–31, 33–4, 36, 39–40, 72, 108 (B)); nos.35, 38 and 58–9 from about 1766–7 are similar. They exhibit great variety of style, subject matter and orchestral treatment, although the common notion that they constituted a distinctly 'experimental' phase is untenable. Their stylistic élan and virtuoso brilliance are attributable to the splendour of the court and the professional players now at Haydn's disposal. One finds works for connoisseurs (nos.6–8, 13, 21–2, 31), others that seek to entertain (nos.9, 16, 33, 36, 72, 108) and still others that combine both stances (nos.34, 39–40). A few present an apotheosis of the chamber symphony: at ease, yet refined and profound (nos. 28–9, 35). Extra-musical aspects are present not only in the *Matin-Midi-Soir* trilogy but also nos.30 ('Alleluja'), 31 ('Hornsignal'), and perhaps 22 ('The Philosopher') and 59 ('Fire', a modern nickname deriving from its supposed origin as incidental music). Although a few symphonies are still in three movements (nos.9, 12, 16, 30), four is the norm at this stage. Concertante scoring is prominent not only in nos.6–8 but nos.9, 13, 14, 16, 31, 36, 72 and 108; a special effect found in this period alone is the use of four horns rather than the usual two (nos.13, 31, 39, 72).

Haydn's symphonies of the years around 1770 (nos.26, 41–9, 52, 65) are widely described as exemplifying his *Sturm und Drang* style; those of 1773–4 (nos.50, 51, 54–7, 60, 64), while less extreme, have many points of contact with it. The most commonly cited feature is the minor mode – of Haydn's ten symphonies in the minor, six fall between 1765 and 1772 – although most works remain in the major, and most of the novel stylistic features are independent of mode. These include remote keys (no.45, 'Farewell', in F♯ minor and major, and no.46 in B major), rhythmic and harmonic complexities, expansion of outward dimensions and harmonic range, rhythmic instability, extremes of dynamics and register, greater technical difficulty, increased use of counterpoint (e.g. in the canonic minuet of no.44, 'Mourning'), musical ideas that seem dynamically potential rather than self-contained, and contrast within themes instead of merely between them. The slow movements and finales become more nearly comparable to the first movements in size and weight; in the former the violins play *con sordino* and the tempo is usually slowed to Adagio. No.26 ('Lamentatione') has religious associations and no.49 ('La passione') may have as well. The programmatic nos.45–6 (they seem to be a pair) are integrated in a through-composed, end-orientated manner not seen again until Beethoven's Fifth Symphony.

From about 1775 (in some respects 1773) to 1781 Haydn again changed his orientation. Symphonies nos.53, 61–3, 66–71 and 73–5 are primarily in a light, even popular style (only no.70 is an exception), perhaps reflecting his resumption of operatic composition in 1773; indeed nos.53, 62, 63 and 73 include adaptations of stage-music, as had nos.50 (1773) and 60 (1774) before them. This stylistic turn has been interpreted as a kind of relaxation, or even as an outright selling-out, but it is better understood as representing the distinct artistic stance of entertainment. They are easy (as Haydn said), but superbly crafted and abound in striking and beautiful passages, not to mention witty and eccentric ones: works of comic genius that approach the *buffa* stage. The slow movements exhibit new formal and stylistic options (the hymn-like no.61, the exquisitely 'popular' theme in no.53, the play of comic and serious in no.68, the ethereal dream in no.62), while the finales adumbrate rondo and hybrid forms. Slow introductions become important about 1779 and begin to create tangible links to the allegros (nos.53, 71 and 73).

During the 1780s Haydn's style changed again, as he began to sell his symphonies abroad, in 'opus' format. Although in many respects nos.76–81 (1782–4) are still 'easy', they include superb movements such as the opening allegro of no.81 and the finales of nos.77 (with its contrapuntal development) and 80 (with its cross-rhythm theme). In nos.78 and 80 Haydn returns to the minor, although from no.80 on he usually ends such movements in the major, and places the entire finale in the major as well. The Paris symphonies (nos.82–7) are the grandest he had yet composed. Nos.87, 83 and 85 (1785) already have a new *esprit*, a combination of learned and popular style, consistency of musical argument and depth of feeling; see the slow movements of nos.83 and 87 and the outer movements of no.85 (the opening Vivace is particularly graceful and harmoniously constructed). In nos.82 and 86 (1786) the trumpets and drums lend added brilliance and the outer movements are on a still larger scale; the Capriccio of no.86 is one of Haydn's most original slow movements. All these features characterize nos.88 and 90–2 as well (no.89 falls off somewhat). Nos.88 and 92 are the best known: the former boasts concentrated, in part contrapuntal outer movements, while the gorgeous Largo theme is set off by entries of the trumpets and drums (withheld from the first movement for this purpose); the latter features an unusually close integration of slow introduction and Allegro, beautiful Adagio, rhythmically intricate trio and Haydn's sprightliest and wittiest finale to date.

Haydn's London symphonies (nos.93–104) crown his career as a symphonic composer. Not only do they outdo the Paris symphonies stylistically, but he produced them in person for rapturous audiences; this interaction stimulated him to ever bolder and more original conceptions. Nos.95–6 (1791) most nearly resemble the preceding symphonies, although no.95 in C minor has a gripping opening movement dominated by a striking unison motto, an ominously terse minuet and a brilliant sonata-fugal finale in C major. Those given in 1792 (nos.93–4, 97–8) respond to Haydn's public: in no.94 the famous outburst in the Andante is actually the least remarkable 'surprise'; the opening Vivace reaches new heights of tonal wit and expansive brilliance, and the concluding sonata-rondo is the first to exhibit the blend of rhythmic vitality, playful surprise, larger scale and underlying cogency of argument that distinguish Haydn's London finales. These last features are found in nos.97–8 as well, along with a new romanticism in the opening movement of no.97 (the breathtaking diminished seventh chord in bar 2, which returns at several keypoints, and the remote flat-side modulations in the recapitulation); in no.98 Haydn composed an extended fortepiano obbligato for himself in the coda of the finale.

The last six symphonies are even more brilliant (clarinets are added, except in no.102); Haydn's determination to conquer new territory with each work is palpable. No.99 is his most elaborate symphonic essay in remote tonal relations; it also features a particularly warm slow movement, with extensive wind writing (much commented on at the time). No.101 ('Clock') has by far the longest minuet and trio Haydn ever composed and a particularly brilliant rondo finale. No.100 ('Military') rapidly became his most popular, owing to the slow movement based on a romance (from the lyre concerto H VIIh:3), overlaid by massive percussion outbursts that audiences found deliciously terrifying. No.102 is the least 'characteristic' of these six, yet one of the greatest; its most remarkable movement is the Adagio (identical in musical substance to that in the Piano Trio H XV:26), in which the exposition is repeated in order to vary the instrumentation, with muted trumpets and drums. No.103 ('Drumroll') offers Haydn's most telling invocation of the sublime in instrumental music, by means of an astonishing double annunciation: first the 'intrada' solo drumroll, then the mysterious bass theme (resembling the 'Dies irae'), which dominates the Allegro as well and, even more astonishingly, interrupts the recapitulation near the end. No.104 begins with a massive dotted motif on the 5th D-A, which some commentators describe as dominating the entire symphony; the first movement is one of Haydn's freest and the finale has greater relative weight than that in any other of the London symphonies.

Besides the symphonies Haydn's orchestral music comprises the six early *Scherzandi* (H II:33–8), a few miscellaneous symphonic movements, overtures and instrumental numbers from operas and oratorios, incidental music, more than 100 minuets (many lost), of which the most important are the magnificent minuets and German dances H IX:11–12 (1792), and four late marches. He also composed numerous concertos, both for melody instruments (many of them lost) and for keyboard. Of the former, the most important are two virtuoso early Esterházy works: the Violin Concerto in C (H VIIa:1) and the massive Cello Concerto in C (H VIIb:1), and two late works: the Concertante (H I:105, 1792) and the Trumpet Concerto (1796), composed for Anton Weidinger's 'keyed' trumpet (a forerunner of the valve trumpet). The six concertos for two *lire organizzate* (H VIIh:1–5; the sixth is lost), commissioned by the King of Naples in 1786, represent a special case; restricted to the keys of C, G and F and by the technical limitations of the instruments, they are Haydn's shortest and most modest concertos, though delightful in every way.

Haydn's three earliest keyboard concertos (1756 to *c*1761) were probably composed for organ, although they were more widely disseminated as harpsichord works; H XVIII:1 in C (?1756) is his earliest surviving large-scale instrumental composition, while no.6 in F is an unusual double concerto for organ or harpsichord and violin. Later came nos.3 in F (probably *c*1770) and 4 in G (probably the early 1770s), both for harpsichord, and no.11, the Piano Concerto in D (*c*1783–4), Haydn's only popular work in this genre. A distinct subgenre comprises the early concertinos (H XIV:11–13, XVIII:F2), not easily distinguished from a group of similar, probably soloistic divertimentos (H XIV:3, 4, 7–10); all are tiny works for harpsichord, violins and bass, mainly in C. Although finely crafted, his keyboard concertos are less original and less popular than his symphonies, perhaps in part because he favours the middle register (except in no.11), eschews both overt and technical display and cantabile writing (except in slow movements), and includes many sequential passages. (These features reflect a particular stylistic orientation, not limitations on Haydn's imagination or his prowess as a performer. The old canard that he was a mediocre keyboard player has long been laid to rest; his statement to Griesinger that 'I was no mean keyboard player and singer' was clearly an understatement, for he continued, 'I could also perform a concerto on the violin'.)

CHAMBER MUSIC WITHOUT KEYBOARD

Haydn's chamber music centres around his 68 string quartets, a genre of which he was more nearly the literal 'father' than the symphony. (The traditional figure of 83 included the spurious op.3, three genuine early works that are not quartets, op.1 no.5 and op.2 nos.3 and 5, and the *Seven Last Words*, but omitted the early H II:6, 'op.0'.) His earlier quartets were composed in three discrete groups separated by long pauses: the ten early works for Baron Fürnberg (in the mid-late 1750s), opp.9, 17 and 20 (*c*1770), and op.33 (1781). Each group offers a different solution to the technical and aesthetic aspects of the genre while cumulatively enlarging the resources of quartet style. The Fürnberg quartets already take the soloistic ensemble for granted, including solo cello without continuo. They belong to the larger class of ensemble divertimentos, with which they share small outward dimensions, prevailing light tone (except in slow movements) and a five-movement pattern, usually fast–minuet–slow–minuet–fast. Even on this small scale, high and subtle art abounds: witness the rhythmic vitality, instrumental dialogue and controlled form of the first movement of op.1 no.1 in B♭; the wide-ranging development and free recapitulation in the first movement of op.2 no.4 in F, and the pathos in its slow movement; and the consummate mastery of op.2 nos.1–2.

Opp.9, 17 and 20 established the four-movement form with two outer fast movements, a slow movement and a minuet (although in this period the minuet usually precedes the slow movement). They also – op.20 in particular – established the larger dimensions, higher aesthetic pretensions and greater emotional range that were to characterize the genre from this point forward. They are important exemplars of Haydn's *Sturm und Drang* manner: four works are in the minor (op.9 no.4, op.17 no.4, op.20 nos.3 and 5); and nos.2, 5 and 6 from op.20 include fugal finales. Op.20 no.2 exhibits a new degree of cyclic integration with its 'luxuriantly' scored opening movement (Tovey, N1929–30), its minor-mode Capriccio slow movement which runs on, *attacca*, to the minuet (which itself mixes major and minor), and the combined light-serious character of the fugue. Op.17 no.5 and op.20 also expand the resources of quartet texture, as in the opening of op.20 no.2, where the cello has the melody, a violin takes the inner part and the viola executes the bass.

In op.33 these extremes are replaced by smaller outward dimensions, a more intimate tone, fewer extremes of expression, subtlety of instrumentation, wit (as in the 'Joke' finale of no.2 in E♭) and a newly popular style (e.g. in no.3 in C, the second group of the first movement, the trio and the finale). Haydn now prefers homophonic, periodic themes rather than irregularly shaped or contrapuntal ones; as a corollary, the phrase rhythm is infinitely variable. The slow movements and finales favour *ABA* and rondo forms rather than sonata form. However, these works are anything other than light or innocent: no.1 in B minor is serious throughout (the understated power of its ambiguous tonal opening has never been surpassed), as are the slow movements of nos. 2 and 5. Op.33 has been taken as marking Haydn's achievement of 'thematische Arbeit' (the flexible exchange of musical functions and development of the motivic material by all the parts within a primarily homophonic texture); although drastically oversimplified, this notion has had great historiographical influence. These quartets' play with the conventions of genre and musical procedure is of unprecedented sophistication; in thus being 'music about music', these quartets were arguably the first modern works.

The appearance of op.33 was the first major event in what was to become the crucial decade for the Viennese string quartet, as Mozart and many other composers joined Haydn in cultivating the genre. Indeed, all the elements of Classical quartet style as it has usually been understood first appeared together in Mozart's set dedicated to Haydn (1782–5). He responded in opp.50, 54/5 and 64 by combining the serious tone and large scale of op.20 with the 'popular' aspects and lightly worn learning of op.33. The minuet now almost invariably appears in third position; the slow movements, in *ABA*, variation or double variation form are more melodic than those in op.33; the finales, usually in sonata or sonata-rondo form, are weightier. Haydn's art is no longer always subtle; the opening of op.50 no.1 in B♭, with its softly pulsating solo cello pedal followed by the dissonant entry of the upper strings high above, is an overt stroke of genius, whose implications he draws out throughout the movement.

Haydn's quartets of the 1790s adopt a demonstratively 'public' style (often miscalled 'orchestral'), usually attributed to his experience in London (op. 71/4 was composed for his second visit there); for example, the fireworks for Salomon in the exposition of op.74 no.1 in C. Without losing his grip on the essentials of quartet style or his sovereign mastery of form, he expands the dimensions still further, incorporating more original themes (the octave leaps in the first movement of op.71 no.2), bolder contrasts, distantly related keys (from G minor to E major in op.74 no.3) etc. Opp.76–7, composed back in Vienna, carry this process still further, to the point of becoming extroverted and at times almost eccentric: see the first movements of op.76 no.2 in D minor, with its obsessive fifths, and of op.76 no.3 in C, with its exuberant ensemble writing and the 'Gypsy' episode in the development, or the almost reckless finales of nos.2, 5 and 6 and op.77. He experimented as well with the organization of the cycle: op.76 nos.1 and 3, though in the major, have finales in the minor (reverting to the major at the end), while nos.5–6 begin with non-sonata movements in moderate tempo (but a fast concluding section), so that the weight of the form rests on their unusual slow movements (the Largo in F♯ of no.5, the tonally wandering Fantasia of no.6).

In his earlier years especially, Haydn composed extensively in other chamber genres. His surviving authenticated ensemble divertimentos (HII) consist of one string quintet (no.2), numerous mixed works including three in nine parts (nos.9, 17 and 20), one each in eight and seven (nos.16 and 8), two sextets for strings and two horns (nos.21–2) and two more for melody instruments (nos.1, 11), as well as at least five works for winds, four sextets (nos.3, 7, 15 and 23) and a tiny piece for two clarinets and two horns (no.14). Most of them exhibit the same five-movement cyclic pattern as the early string quartets, with the difference that contrasts in instrumentation become a basis of style, for example in reduced scorings in trios and slow movements or extended soloistic passages. Although some of the mixed works (nos. 1–2, 9, 11 and 20) are among the earliest and are on average the least compelling, the slightly later nos.8, 16–17 and 21–2 are on the same high level as the quartets. The wind band works seem to date from about 1760–61; they are even smaller in scale but unfailingly masterful.

By contrast, the 21 authenticated string trios (HV:1–21, by 1765) are works for connoisseurs in 'high' style, difficult for player and listener alike, in a wide range of keys (three in E, one in B, two even in B minor). All are scored for two violins and (presumably) cello except no.8 (violin and viola) and are thus related to the trio sonata tradition, although the first violin dominates more than it participates in dialogue. They are in three movements (except no.7, in two), with a bewildering variety of cyclic patterns; many begin with a slow movement and most include a minuet. The 126 baryton trios (HXI; c1762–75) are similar in that they are music for a (particular) connoisseur and always in three movements with a minuet (except no.97 with seven: 'fatto per la felicissima nascita di S.A.S. Prencipe Estorhazi'). Although the baryton takes the leading role, they include much dialogue and 'thematische Arbeit'; three late works (nos.97, 101 and 114) include fugues. They are intimate music, modest rather than ambitious, with a narrow range of keys (dictated by the baryton's technical limitations); Haydn's ability to fashion genuine art within such restricted conditions is remarkable.

In the middle and late 1770s Haydn's production of chamber music fell off markedly. One last group of baryton works comprises the important octets HX:1–6 (mid-1770s); they are richer in scoring and on a larger scale than the trios. The six violin-viola duets (HVI:1–6) are likewise from the mid-1770s; the violin dominates and the style seems somewhat old-fashioned. The six string trios from the early 1780s (HIV:6–11) and the four flute trios from London (HIV:1–4, 1794–5) are amateurs' music, with small dimensions, simple textures and restriction to two or three movements. By contrast, the eight lyre notturnos for the King of Naples (HII:25–32, 1788–90), of unfailing charm and true 'chamber' disposition, offer a wonderful synthesis of play and art.

KEYBOARD MUSIC

Haydn's keyboard works comprise solo sonatas (HXVI), trios (HXV) and quartet-divertimentos (HXIV). In 18th-century thought and practice these constituted a single, loosely defined genre, destined primarily for private performance and orientated on the topic of sentiment, seen as the natural expressive mode for music performed solely or primarily by an individual at the keyboard; indeed he often adopts a selfconsciously improvisatory style, especially after 1780. During the 1760s they were apparently composed for the harpsichord. The first clear (albeit indirect) evidence of composition for the fortepiano (or possibly clavichord) is found in the highly expressive Sonata no.20 (1771), with mannered dynamic marks. Nevertheless, most works from the 1770s may have been conceived for the harpsichord or neutrally for both instruments. Beginning in the early 1780s, and decisively from the late 1780s on, Haydn composed for the fortepiano. Many of his keyboard works were composed for ladies, whether students in his early years, the Auenbrugger sisters around 1780, or intimates such as Mme Genzinger, Mrs Schroeter and Therese Jansen. The majority are in three movements: either fast–slow–fast, or a fast movement, slow movement and minuet in various permutations. Two-movement works are also common, often slow–fast; numerous slow movements in penultimate position are run on, *attacca*, to the finale. Even in the 1780s and 90s many works end with an outwardly modest movement such as a Tempo di menuetto, a set of variations or a simple rondo. Neither the two- and three- movement cyclic patterns nor the modest finales were 'conservative' or 'immature', as has been claimed; they are as finely wrought as quartet finales and exemplify the prevailing generic orientation of intimacy.

Haydn's early keyboard works are both serious and *galant*. The trios HXV:f1 in F minor and 1 in G minor and Sonata no.2 in B♭ (with its astonishing Largo) are more intellectually difficult and stylistically uncompromising than all the early quartets and most of the early symphonies; many works are small and unpretentious and were presumably written for students and amateurs. At least 12 weighty connoisseurs' sonatas originated in the late 1760s and early 1770s, including nos.19, 20, 45, 46 and seven lost works. Two sets in mixed style followed, nos. 21–6 (1773) and 27–32 (1774–6); they include serious works such as the boldly formed nos.22 in E and 26 in A, the passionate no.32 in B minor and the through-composed no.30 in A, as well as numerous lighter works, especially in the latter set. In 1780 followed his first publication with Artaria, the heterogeneous nos.35–9 and 20, including the 'easy' no.35 in C, the virtuoso no.37 in D and the serious no.36 in C♯ minor. The three modestly scaled sonatas nos.40–42 (published 1784) are miracles of popular appeal allied with high art, especially no.40 in G. Except for no.51 in D for Mrs Schroeter, Haydn's last five sonatas eschew any pretence of modesty. In the late 1780s he composed no.48 in C, with a fantasy-like slow variation movement and a dashing sonata-rondo finale, and the intimate no.49 in E♭ for Mme Genzinger; its brilliant first movement has an unusually long coda and the *ABA* Adagio is richly expressive, with continual variations of the theme. From London come two virtuoso sonatas for Jansen: nos.50 in C and 52 in E♭. The former features a remarkable first movement which, though in sonata form, is based on continual variation of a basic motif; the latter is on the largest scale throughout and features a slow movement in the remote key of E major (a tonal relation adumbrated in the development of the first movement and wittily 'cancelled' at the beginning of the finale).

Of Haydn's few keyboard works outside the sonatas, the most important are two capriccios – *Acht Sauschneider müssen sein* in G (HXVII:1, 1765), a variation rondo with an immense tonal rnge, and HXVII:4 in C (1789), another tonally wide-ranging work with elements of sonata rondo form, perhaps stimulated by a Fantasia from C.P.E. Bach's sixth collection of *Clavier-Sonaten ... für Kenner und Liebhaber* (1787) – and the F minor Variations for piano HXVII:6 (1793), arguably Haydn's most original and concentrated double-variation movement, with a coda (added in revision) of Beethovenian power.

Haydn's piano trios have been undervalued, in part because of the great distance between their original generic identity and today's conceptions. 18th-century keyboard trios (like violin sonatas) were understood as 'accompanied sonatas': the keyboard dominates, the cello mainly doubles the left hand of the piano in a pitch-class sense, and even the violin is generally more accompanimental than soloistic, although it often receives sustained melodies in second themes, slow movements, minuet trios and rondo episodes. Nevertheless the strings are essential, for integration of the texture, tone colour and rhythmic definition. The effort to hear Haydn's 27 late piano trios (HXV:5–31,1784–96) with 18th-century ears is worth making: after the quartets they comprise his largest and greatest corpus of chamber music. No.12 in E minor (1788–9) has an opening movement of astonishing seriousness with vast expansions towards the end, while the beautiful siciliano slow movement and the ebullient rondo finale are both in E major. No.14 in A♭ (1789–90) includes his first slow movement in a remote key (E major, or ♭VI, adumbrated by B major in the development of the first movement). From London, nos.24–6 (1794–5), dedicated to Schroeter, include no.25 (with the famous 'Gypsy Rondo') and no.26 in the special key of F♯ minor: following a concentrated and brooding Allegro and a gorgeous Adagio in F♯ major (identical in substance with the Adagio of Symphony no.102), the minuet-finale is anything other than anticlimactic. It begins dissonantly on a dominant ninth and this instability is maintained throughout:

there is no tonic cadence until the very last bar of the *A* section, just before the double bar, and in the reprise even this cadence is deceptive, leading to a substantial coda. Rosen praises its 'intimate gravity ... a melancholy so intense it is indistinguishable from the tragic', while Landon conjectures that the work may represent Haydn's farewell to Mrs Schroeter (the key is suggestive). By contrast, nos.27 in C and 29 in E♭ (1795–6), dedicated to Jansen, are difficult and extroverted; no.29 is particularly original in construction, and both have rollicking finales that outdo any earlier ones.

9. HAYDN'S CAREER

Haydn's career never stimulated a paradigmatic narrative comparable to that of Beethoven's three periods. To be sure, decisive breaks occurred in 1761 (his move to the Esterházy court), 1790 (to London) and 1795 (back to Vienna); the periods 1750–61, 1791–5 and 1795–1802 are distinctive regarding both the conditions of his life and his compositional activity. However, the first and last of these are brief in proportion to his career as a whole and cannot bear the weight that 'early' and 'late' do in Beethoven's case. Furthermore, in any such reading Haydn's 30 years at the Esterházy court remain a long, uninterpreted 'middle'. Its only major dividing-points that affected both his life circumstances and his compositional orientation were 1766, when he became full Kapellmeister, 1776, when he became responsible for the court opera, and 1779, when he negotiated his independence as composer of instrumental music. Hence except for 1761–5 the Esterházy years seem best understood in terms of a series of overlapping phases, each defined by different criteria.

In the 20th century too much was made of the supposedly evolutionary aspects of Haydn's career, in part because of its association with the notion of the rise of 'Classical style'. This led to a threefold periodization after all, but one modelled mechanically on the traditional interpretation of artists' careers: apprentice – journeyman – master. In Haydn's case this took the form: immaturity/composition within existing style – experimentation/searching for a new style – maturity/'Classical style'; the last was assumed to be his overriding stylistic 'goal', which he finally 'achieved'. The oldest and most persistent of these interpretations associated Classical style with 'thematische Arbeit' and the string quartets op.33 of 1781. Another proposed a double progression: towards a first highpoint with his *Sturm und Drang* manner around 1768–72, and a second one with the Paris symphonies and the *Seven Last Words* of 1785–6. These notions are not facts, however, but constructions, placed in the service of stylistic narratives of the 'per aspera ad astra' type, more ideologically focussed and psychologically reassuring than explanatory. To be sure, other things equal, a later work of Haydn will be more complex and concentrated than an earlier one; indeed his music often became 'more so' within a single genre over a few brief years; for example the string quartets opp.9–17–20 or the London symphonies. And he certainly experimented compositionally, as is clear from his own account of 'becoming original'. But even his earliest music was never in any intrinsic sense immature, and he continued to experiment – successfully – throughout his career. From about 1755 on, Haydn's music was technically masterful, generically appropriate and rhetorically convincing; every one of his works is best appreciated today in terms of these three modes of understanding, applied in concert.

WORKS

Editions: *Joseph Haydns Werke*, ed. E. Mandyczewski and others, 10 vols. (Leipzig, 1907–33) [M]
            Joseph Haydn: Kritische Gesamtausgabe, ed. J.P. Larsen, 4 vols. (Boston, Leipzig and Vienna, 1950–51) [L]
            Joseph Haydn: Werke, ed. J. Haydn-Institut, Cologne, dir. J.P. Larsen (1958–61), G. Feder (1962–90) and others
               (1990–), 78 vols. (Munich, 1958–) [HW]
            Joseph Haydn: Kritische Ausgabe sämtlicher Symphonien, i–xii, ed. H.C.R. Landon, Philharmonia series (Vienna,
1965–8)   [P]
            Diletto musicale, ed. H.C.R. Landon unless otherwise stated (Vienna, 1959–) [D]
            (for editions of specific genres, see notes at head of relevant sections)
Catalogue: A. van Hoboken: *Joseph Haydn: Thematisch-bibliographisches Werkverzeichnis*, i: *Instrumentalwerke*; ii: *Vokalwerke*; iii:
            Register, Addenda [Add.] und Corrigenda (Mainz, 1957–78) [H]

    1. VOCAL: A. Masses. B. Miscellaneous sacred. C. Oratorios and similar works. D. Secular cantatas, choruses. E. Dramatic.
F. Secular vocal with orchestra. G. Solo songs with keyboard. H. Miscellaneous vocal works with keyboard. I. Canons.

    2. INSTRUMENTAL: J. Symphonies. K. Miscellaneous orchestral. L. Dances, marches for orchestra/military band. M. Concertos
for string or wind instruments. N. Divertimentos etc. for 4+ string and/or wind instruments. O. String quartets. P. String trios
(divertimentos). Q. Baryton trios (divertimentos). R. Works for 1–2 barytons. S. Miscellaneous chamber music for 2–3 string
and/or wind instruments. T. Works for 2 lire organizzate. U. Keyboard concertos/concertinos/divertimentos. V. Keyboard trios.
W. Keyboard sonatas. X. Miscellaneous keyboard works. Y. Works for flute clock.

    3. FOLKSONG ARRANGEMENTS: Z. Arrangements of British folksongs.

Authentication symbols:

A	– autograph, i.e. written and signed by Haydn or marked 'In nomine Domini', 'laus Deo' (or similarly) by him	HE	– MS copy from Haydn's estate
		HL	– autograph entry in Haydn's list of librettos
		HV	– thematic entry in Haydn-Verzeichnis, 1805
C	– MS copy by one of Haydn's copyists: Anon.11, 12, 30, 48, 63 (nos. from Bartha–Somfai)	H 1799– 1803	– verified/rev. by Haydn in those years according to C.F. Pohl's papers, *A-Wgm* (based on lost documents in Breitkopf archives)
Dies	– his book on Haydn, 1810		
E	– MS copy by Johann Elssler		
EK	– entry in Haydn's Entwurf-Katalog, *c*1765–	JE	– MS copy by Joseph Elssler sr
F	– MS copy by one of 3 earliest copyists of H-KE Fürnberg collection; found in various archives	OE	– original edition, published by Haydn or authorized by him
Gr	– Griesinger's book on Haydn, 1810, or his letters to Breitkopf & Härtel	RC	– MS copy rev. Haydn
		SC	– MS copy signed by Haydn
HC	– entry in non-thematic list of Haydn's music collection, *c*1807	Sk	– sketch by Haydn
		u	– unsigned

1766 = composed 1766; [1766] = year of composition 1766 not documented; –1766 = composed by 1766; –?1766 = possibly
composed by 1766

— signifies the absence of the work from the category concerned; i.e. not in H, not authenticated, not pubd etc.

Items are numbered chronologically (as far as possible) within each category (except in section Z); these numbers are always
shown in italics and are used for cross-references between sections (e.g. E 23).

Where not specified, bn may often double the bass part.

Instrumental parts that are doubtful or are later additions (sometimes by Haydn himself) are parenthesized or given a question
mark.

Numbers in the right-hand column denote references in the text.

1: VOCAL

A: MASSES

No.	XXII	Title, key	Forces	Date	Authentication	Edition	Remarks	
1a	3	Missa 'Rorate coeli desuper', G		?	EK, HV	?	?lost/?identical with no.1b	5, 8, 54
1b	ii, 73	Mass, G	4vv, 2 vn, bc (org)	–1779	?	HW xxiii/1a, 207	by G. Reutter jr/Arbesser/Haydn	
2	1	Missa brevis, F	2 S, 4vv, 2 vn, bc (org)	?1749	SC, ?EK	L xxiii/1, 1; HW xxiii/1a, 1	wind and timp pts added by Haydn (or Heidenreich), 1805/6	5, 16, 45, 54
3	5	Missa Cellensis in honorem BVM (Cäcilienmesse), C	S, A, T, B, 4vv, 2 ob, 2 bn, ? 2 hn, 2 tpt, timp, str, bc (org)	1766 [–?1773]	EK, A (frags.)	L xxiii/1, 105; HW xxiii/1a, 29	doubtful hn pts in Bs only	16, 54, 56
4	2	Missa Sunt bona mixta malis, d	4vv, bc (org)	1768	EK, A (frag.)	HW xxiii/1b, 166	Ky and 1st section of Gl extant	16, 45, 54
5	4	Missa in honorem BVM (Missa Sancti Josephi; Grosse Orgelsolomesse), Eb	S, A, T, B, 4vv, 2 eng hn, 2 hn, (2 tpt, timp), 2 vn, bc, org obbl	–1774 [?c1768–9]	EK, A (frags., u)	L xxiii/1, 24; HW xxiii/1b, 1	tpt and timp in authentic MS copy (?E), H-Gk	16, 54
6	6	Missa Sancti Nicolai (Nikolaimesse; 6/4-Takt-Messe), G	S, A, T, B, 4vv, 2 ob, 2 hn, str, bc (org)	1772	EK, A	L xxiii/1, 270; HW xxiii/1b, 105	in HV as Missa Sr Josephi; cf no.5; tpts and timp in authentic MS copy (E), 1802	16, 54
7	7	Missa brevis Sancti Joannis de Deo (Kleine Orgelmesse), Bb	S, 4vv, 2 vn, bc, org obbl	–1778 [?c1773–7]	EK, A	HW xxiii/2, 1	see also HW xxiii/2, 247	16, 54
8	8	Missa Cellensis (Mariazellermesse), C	S, A, T, B, 4vv, 2 ob, bn, 2 tpt, timp, str, bc (org)	1782	A	HW xxiii/2, 20	Bs uses aria from Il mondo della luna (E 17)	5, 26, 54
9	10	Missa Sancti Bernardi von Offida (Heiligmesse), Bb	S, A, T, B, 4vv, 2 ob, 2 cl, 2 bn, ? 2 hn, 2 tpt, timp, str, bc (org)	1796	A, Sk	HW xxiii/2, 166	= Missa Sr Ofridi in EK; see also HW xxiii/2, 240, 242; cf I b, 44	38, 54, 55
10	9	Missa in tempore belli (Kriegsmesse; Paukenmesse), C	S, A, T, B, 4vv, ?fl, 2 ob, 2 cl, 2 bn, 2 hn, 2 tpt, timp, str, bc (org)	1796	A	HW xxiii/2, 89	perf. Vienna, 26 Dec 1796; see also HW xxiii/2, 237	38, 51, 54, 55
11	11	Missa (Nelsonmesse; Imperial Mass; Coronation Mass), d	S, A, T, B, 4vv, 3 tpt, timp, str, bc, org obbl	10 July–31 Aug 1798	A	HW xxiii/3, 1	= Missa in angustiis in EK; perf. ?Eisenstadt, 23 Sept 1798; org pt transcr. for wind insts ?by J.N. Fuchs	38, 41, 54, 55
12	12	Missa (Theresienmesse), Bb	S, A, T, B, 4vv, 2 cl, (bn), 2 tpt, timp, str, bc (org)	1799	A	HW xxiii/3, 140		38, 54
13	13	Missa (Schöpfungsmesse), Bb	S, A, T, B, 4vv, 2 ob, 2 cl, 2 bn, 2 hn, 2 tpt, timp, str, bc, org (obbl in Et incarnatus)	11 July–11 Sept 1801	A	HW xxiii/4; facs. (Munich, 1957)	perf. Eisenstadt, 13 Sept 1801; Gl quotes duet from The Creation; see also HW xxiii/4, 204	38, 51, 54
14	14	Missa (Harmoniemesse), Bb	S, A, T, B, 4vv, fl, 2 ob, 2 cl, 2 bn, 2 hn, 2 tpt, timp, str, bc (org)	1802	A	HW xxiii/5	perf. Eisenstadt, 8 Sept 1802	38, 42, 54

Note: over 100 spurious masses listed in Hoboken; composers of some identified by MacIntyre (H1982)

B: MISCELLANEOUS SACRED

No.	H	Title, key	Forces	Date	Authentication	Edition	Remarks	
1	XXIIIc:5	Lauda Sion (Hymnus/Motetto de venerabili sacramento), i–iv, C	4vv, 2 ob, 2 tpt, str, bc (org)	–1776 [?c1750]	?EK	(Vienna and Munich, ?996)	also with Salve regina text	5, 8, 56
2	XXIIIb:3*	Ave regina, A	S, 4vv, 2 vn, bc (org)	–1763 [?c1750–59]	—	(Augsburg 1970)	also with Salve regina text	
3	XXIIIb:1	Salve regina, E	S, 4vv, 2 vn, bc (org)	?1756	EK, A	(Vienna and Munich, ?900)	date on autograph added later	9, 1?, 56
4	XXIIIa:4*	Quis stellae radius (motet), C	S, 4vv, ? 2 tpt, ?timp, str, bc (org)	?1762	SC	—	cant.; also with other texts, incl. Quae admiranda lux; for ?secular origin, see Becker-Glauch (J1970)	
5	XXIIIc:1	Te Deum, C	S, A, T, B, 4vv, timp, 2 vn, bc (org)	–1765 [?1762–3]	?EK	(Vienna and Munich, 1966)	also attrib. M. Haydn	14, 56
6	XXIIIa:3	Ens aeternum (off/motet/hymn), G	4vv, str, bc (org)	–1772 [?1761–9]	HV	(Leipzig, 1813)	also with text Walte gnädig, with addl. 2 ob, 2 tpt, timp	
7	XXIIIa:2	Animae Deo gratae (off/motet), C	2 S, T, 4vv, 2 ob, 2 tpt, timp, str/? 2 vn, bc (org)	–1776 [?1761–9]	HV	—	also attrib. M. Haydn; also with text Agite properate	
8	XXIIIc:4	Lauda Sion (Responsoria de venerabili [sacramento]), i–iv, Bb, d, A, Eb	S, A, T, B (? in chorus), ? 2 hn, 2 vn, bc (org)	?c1765–9	EK	(Munich, 1965) (entitled Hymnus)	MS copy as Quatuor Stationes pro Festo Corporis Christi	56
9	XXIIIc:3	Alleluia, G	S, A, 4vv, str, bc (org)	–1771 [?1768–9]	A	facs. (Eisenstadt, 1976)	in MS copies always fo lowing Dictamina mea (appx B.1, 3)	
10	XXIIId:3*	Herst Nachbä (Cantilena pro adventu/ Pastorella), D	S ? 2 hn, str, bc (org)	?c1768–70	EK	(Altötting, 1975)	also with other texts, incl. Jesu redemptor omnium	
11	XXIIIb:2	Salve regina, g	S, A, T, B, str, bc: org obol	1771	EK, A	(Vienna and Munich, 1964)	1770 incorrect reading	16, 45 50 58
12	XXIIIb:4*	Salve regina, Eb	S, A, T, B (?solo vv), str, bc (org)	–1773	—	(Augsburg, 1959)	?doubtful (see Landon, i A1980, p.157); B in edn ?not orig.	
13	XXIIId:1	Ein' Magd, ein' Dienerin (Cantilena/ Aria pro adventu), A	S, ? 2 ob, ? 2 hn, str, bc (org)	?c1770–75	EK	(London, 1957)		
14	XXIIId:2	Mutter Gottes, mir erlaube (Cantilena/ Aria pro adventu), G	S, A, 2 vn, bc (org)	?c1775	?EK	—		
15	XXIIIa:1	Non nobis, Domine (Ps cxiii:9) (off in stile a cappella), d	4vv, bc (org)	–1786 [?c1768]	EK	(St Louis, 1960); (Vienna and Munich, 1978)	for date see Haydn Yearbook 1992, 168	56
16	XXIIIb:1*	Libera me, Domine, d	S, A, T, B (? in chorus), 2 vn, bc (org)	?c1777–90	A (u pts)	(Salzburg, 1969)	? only copied, ? not by Haydn	

No.	H	Title, key	Forces	Authentication	Date	Edition	Remarks	
17– 22	ii, 181	6 English Psalms (J. Merrick, rev. W.D. Tattersall): 17 How oft, instinct with warmth divine, F (Ps xxvi.5–8); 18 Blest be the name of Jacob's God, E♭ (Ps xxxi.21–4); 19 Maker of all! be Thou my guard, D (Ps xli.12–16); 20 The Lord, th' almighty Monarch, spake, C (Ps l.1–6); 21 Long life shall Israel's king behold, E♭ (Ps lxi.6–8); 22 O let me in th' accepted hour, A (Ps lxix.13–17)	2 S, B	Haydn's 3rd London notebook; RC (no.17)	[1794/5]	(Kassel, 1978)	no.22 uses canzonetta Pleasing Pain (G 29); for MS of no.17 see Haydn Society Journal of Great Britain, xv (1995)	56
23	XXIIIc:2	Te Deum, C	4vv, fl, 2 ob, 2 bn, (2 hn), 3 tpt, timp, str, bc (org)	RC	–Oct 1800	(Vienna and Munich, 1959) (with addl 3 trbn); (Oxford, 1992)	for Empress Maria Theresa	41, 53, 56

Note: Stabat mater, see Group C; The Ten Commandments, see Group I; Ave Maria, mentioned in Elssler, *Haydn's vollendete Compositionen* (MS, *A-Sm*), not identified

Appendix B.1: Selected adaptations and arrangements (authorship uncertain, but Haydn's approval probable in most cases)

No.	Becker-Glaub (J1970)	Title	Edition	Original version	Remarks
1	—	Audi clamorem nostrum (off)	—	final chorus in 1st pt of Il ritorno di Tobia (C 3)	Pohl (A1882), B/m/13; also with other texts
2	B/6/c	Concertantes jugiter (off)	HW xxvii/2, 122	aria Si obtrudat in Applausus (C 2)	
3	B/6/b	Dictamina mea (off/motet)	HW xxvii/2, 68	duetto in Applausus (C 2), combined with Alleluia (B 9)	edn without Alleluia
4	B/8	Insanae et vanae curae (Der Sturm) (off/motet/grad)	(Leipzig, 1809)	chorus Svanisce in un momento in Il ritorno di Tobia (C 3)	not later than 1798; authenticated by E; also with texts Des Staubes eitle Sorgen, Distraught with care and anguish
5	B/7	Maria, die reine (Aria pro adventu)	—	aria of Baucis in Philemon und Baucis (E 12)	
6	B/6/d	O Jesu, te invocamus (off/hymn)	HW xxvii/2, 170	final chorus in Applausus (C 2)	also with text Allmächt'ger, Preis dir und Ehre!
7	—	Plausus honores date (off/motet)	—	final chorus in Da qual gioia (D4)	with orch introduction based on preceding recit
8	B/6/a	Quae res admiranda ... Christus coeli atria (off/motet)	HW xxvii/2, 4, 18		1st recit and qt in Applausus (C 2)
9	B/3	Vicisti, heros ... Justus ut palma	HW xxv/1, 154	recit and aria of Leopoldo in Marchese (E 3)	
10	hXXXIc: 1	Vias tuas Domine (grad), C, 4vv, bc	—	?	by unknown composer, 1576, ? ed. Haydn

Appendix B.2: Selected works attributed to Haydn

No.	H	Title, key	Forces	Earliest reference	Edition	Remarks
1	XXIIIa:5*	Ad aras convolate (grad/off), G	4vv, ? 2 ob, ? 2 trbn, str, bc (org)	1794	—	probably not authentic
2	XXIIIb:E1	Alma redemptoris mater, E	4vv, bc (org)	—	(Vienna and Graz, 1916)	probably not authentic
3	XXIIIa:8*	Ardentes Seraphini (off), A	2 S, str, bc (org)	1765	—	doubtful
4	XXIIIb:6*	Ave regina, F	4vv, ? 2 tpt, ?timp, 2 vn, bc (org)	1782	—	doubtful
5	—	Ego virtus gratitudo (aria), C	S, 2 ob, 2 tpt, timp, str, bc (org)	1772	—	?authentic; for ?secular origin, see Becker-Glauch (J1970), no.B/2
6	XXIIId:G1	Ei wer härt' ihm das Ding gedenkt (Pastorella, aria), G	S, 2 vn, bc (org)	1764	(Altötting, 1975)	?authentic; ? also attrib. (J.A.) Stephan and M. Haydn
7	XXIIIc:6*	Lauda Sion (Aria de venerabili [sacramento]), F	A, 2 fl str, bc (org)	1787?	—	probably not authentic; orig. without author's name
8	XXIIIc:C2	Litaniae de BVM, C	S, A, T, B, 4vv, fl., ? 2 ob, ? 2 tpt, ?timp, 2 vn, bc org obbl	1776	(Vienna and Munich, 1960)	several versions; probably by J. Heyda (Hayda, Haida; c1740–1806); also attrib. M. Haydn
9	XXIIIa:C7	Magna coeli domina (Moretto de Beata, aria), C	B, ? 2 tpt, ?timp, str, bc (org)	—	—	?authentic; for ?secular origin, see Becker-Glauch (J1970)
10	—	Maria Jungfrau rein (Aria pro adventu), G	S, 2 vn, bc (org)	—	—	?authentic; for ?secular origin, see Becker-Glauch (J1970), no.B/5
11	XXIIIa:6*	Salus et gloria (off/motet), C	4vv, 2 tpt, timp, 2 vn, bc (org)	1779	(Augsburg, 1959)	probably by L. Hofmann
12	XXIIIb:5*	Salve regina, G	S, A, 2 vn, bc (org)	1766	—	probably by J. Heyda
13	XXIIIa:7*	Super flumina Babylonis (Ps cxxxvi) (motet), C Veni tandem expectactus, see appx E1 7	A, 4vv, 2 tpt, timp, str, bc (org)	1772	—	probably by Vanhal
14	—	Was meine matre Brust bekränket (Aria pro adventu), G	T, 2 vn bc (org)	—	—	?authentic; MS 'Hayden' in CZ-*Pnm* (Kuks)
15	XXIIIa:G9	O coelitum beati (motet/aria), G	S, str, bc (org)	—	(Cardiff, 1984)	?authentic; for ?secular origin, see preface to edn; in one source with Alleluia, C

Note: for chorus, D, Sit laus plena, sit sonora (text from Lauda Sion) and for recit and aria Quid hostem times, see Landon (A1980)

C: ORATORIOS AND SIMILAR WORKS

No.	H	Title, poet		Date	Authentication	Edition	Remarks	
1	XXbis	Stabat mater (hymn)	S, A, T, B, 4vv, 2 ob/eng hn, str, bc (org/hpd)	1767	EK	HW xxii/1	listed as orat in HV; also other texts, incl. Weint ihr Augen and Traurer Seelen; more insts added Neukomm, 1803 (HW xxii/1, 111)	16, 17, 45, 50, 55
2	XXIVa:6	Applausus (Jubilaeum Virtutis Palatium) (allegorical orat/cant.)	S, A, T, 2 B, 2 ob, bn, 2 hn/tpt, timp, str, bc, hpd obbl	[–4 April] 1768	EK, A	HW xxvii/2	perf. Zwettl, 17 April 1768	53, 17, 61
3	XXI:1	Il ritorno di Tobia (orat, 2 pts, G.G. Boccherini)	2 S, A, T, B, 4vv, 2 fl, 2 ob, 2 eng hn, 2 bn, 4 hn, 2 tpt, 2 trbn, timp, str, bc (hpd)	[1774–5]	A (pt ii; u, and no.13c), C	HW xxviii/1 (I, II)	perf. Vienna, 2, 4 April 1775; rev. and choruses Svanisce in un momento and Ah gran Dio! added 1784; both versions of acc. recits in HW; embellished versions and cadenzas in H7b, 12b dubious; rev. Neukomm with Haydn's permission, 1806; ov., cf K 9	5, 17, 27, 28, 56
4	XX/2	Die Sieben letzten Worte unseres Erlösers am Kreuze (The Seven Last Words) (? J. Friebert, rev. G. van Swieten)	S, A, T, B, 4vv, 2 fl, 2 ob, 2 cl, 2 bn, dbn, 2 hn, 2 tpt, 2 trbn, timp, str	[1795–6]	A (u, partly in copyist's hand), OE	HW xxviii/2	for Haydn's orig. orch version, see K 11; uses also J. Friebert's vocal arr.; text partly uses K.W. Ramler: Der Tod Jesu; perf. Vienna, 26, 27 March 1796	26, 27, 35, 38, 43, 45, 56, 73
5	XXI:2	Die Schöpfung (The Creation) (orat, 3 pts, van Swieten, after unidentified 'Lidley', after Bible and Milton: Paradise Lost)	S, T, B, 4vv, 3 fl, 2 ob, 2 cl, 2 bn, dbn, 2 hn, 2 tpt, 3 trbn, timp, str, bc (hpd)	1796–8	OE, RC, Sk, A (dbn, trbn pts only, u)	M 16/V, ed. A.P. Brown (London, 1995)	perf. Vienna, 29, 30 April 1798; several sketches in Landon, iv (A1977), 357	31, 33, 38, 39, 40, 41, 42, 43, 44, 45, 50, 53, 56–57

No.	H	Title, poet	Forces	Date	Authentication	Edition	Remarks	
6	XXI:3	Die Jahreszeiten (The Seasons) (4 pts, van Swieten, after J. Thomson)	S, T, B, 4vv, 2 fl, 2 ob, 2 cl, 2 bn, dbn, 4 hn, 3 tpt, 3 trbn, timp, perc, str, bc (hpd)	1799–1801	OE, RC, Sk	M 16/VI–VII	text incl. 2 songs by C.F. Weisse and G.A. Bürger; perf. Vienna, 24 April 1801; aria Schon eilet quotes Andante from J 94; for dbn, see Landon, v (A1977), 132	38, 40, 41, 45, 46, 56, 57–8

Note: Die Erlösung mentioned in HL not verified; ?identical with no.4 or with spurious arr. Der Versöhnungstod (H Anh.XXIVa:1)

D: SECULAR CANTATAS, CHORUSES

No.	H	Title	Forces	Date	Authentication	Edition	Remarks	
1	XXIVa:1	Vivan gl'illustri sposi (cant.)	?	–10 Jan 1763	EK	—	lost; for Anton Esterházy's wedding; in EK as Coro 1	61
2	XXIVa:2	Destatevi o miei fidi (canc.)	2 S, T, 4vv, 2 ob, 2 hn, str, bc (hpd)	[–26 Dec] 1763	A, ?EK	—	for Nikolaus Esterházy's nameday; ? mixed with other pieces; autograph, PL-Kj	14, 61
3	XXIVa:4	Qual dubbio ormai (cant.)	S, 4vv, 2 ob, 2 hn, str, bc, hpd obbl	[–26 Dec] 1764	A, ?EK	D 200 (1982)	for Nikolaus Esterházy's nameday; for autograph of final chorus, see Fine Music Manuscripts (B1990)	14, 61
4	XXIVa:3	Da qual gioia improvvisa (cant.)	S, A, 4vv, fl, 2 ob, bn, 2 hn str, bc, hpd obbl	?1764	A, ?EK	—	for Nikolaus Esterházy's return from Frankfurt; ?inc.; autograph, PL-Kj; see appx B.1, 7	14, 61
5	XXIVa:5	Dei clementi (cant.)	?	?	EK	—	lost; for Nikolaus Esterházy's convalescence; in EK as Coro 3	14, 61
6	XXIVa:3	Al tuo arrivo felice (cant.)	?	?1767	EK	—	lost; for Nikolaus Esterházy's return from Paris; in EK as Coro 2; in H confused with no.4	14, 51
7	ii, 433	Su cantiamo, su beviamo (chorus)	S, 3vv, fl, 2 ob, 2 hn, 2 tpt, timp, str	?1791	A (u)	—	adapted from final chorus of Orlando paladino (E 22); cf E 24	35, 61
8a	XXIVa:8	The Storm: Hark! The wild uproar of the winds (P. Pindar, madrigal)	S, A, T, B (?and 4vv), 2 fl, 2 ob, 2 bn, str	[?–24 Feb] 1792	A	ed. F. Szekeres, D 316 (1969)		
8b	ii, 194	Der Sturm: Hört! Die Winde furchtbar heulen (? van Swieten, chorus)	S, A, T, B, 4vv, 2 fl, 2 ob, 2 cl, 2 bn, 2 hn, 2 tpt, 2 trbn, t.mp, str	–1798 [?1793]	A (u, partly in copyist's hand), SC	(Leipzig, 1802)	Ger. trans. of no.8a; also as La tempesta with It. text	

No.	H	Title	Forces	Date	Authentication	Edition	Remarks	
9	XXIVa:9	Nor can I think ... Thy great endeavours (from Klareamontos: [Invocation] Neptune to the Commonwealth of England)	B, 4vv, fl, 2 ob, 2 cl, 2 bn, 2 hn, 2 tpt, timp, str	?1794	A (u)	D 90 (1990)	aria and chorus from inc. cant.; text from prefatorial poem to Selden's Mare clausum	38, 43, 62
10	—	Song with orch	?	1791–5	Gr, Dies	—	lost/unidentified	
11	XXVIa:43	Gott erhalte Franz den Kaiser (L.L. Haschka, Volkslied)	1v, fl, 2 ob, 2 bn, 2 hn, 2 tpt, timp, str	1797	A	(London, 1977) (see Landon, iv, A1977, p.279); facs. (Graz, 1982)	for orig. version, see G 43; perf. 12 Feb 1797	

Note: Quis stellae radius, see Group B

Appendix D.1: Arrangement

No.	H	Title, poet	Forces	Date	Authentication	Edition	Remarks
1	—	God save the King	?	1791–5	Gr, Dies	—	lost

Appendix D.2: Doubtful and spurious works

No.	H	Title	Forces	Edition	Remarks
1	XXIVa:11*	Die Erwählung eines Kapellmeisters (cant.)		ed. F. Szekeres, D 374 (1970)	MSS not authentic
2	Ia:D4	D'onora al piede pongansi (ov., chorus)		—	frag. without author's name; not a Haydn autograph
3	XXIVa:D2	Inimica mihi semper sydera (Applausus)		—	frag. of cant. without author's name; not a Haydn autograph

E: DRAMATIC

No.	H	Title, librettist	Forces	Date	Authentication	Edition	Remarks	
1a	XXIXb:1a	Der krumme Teufel (Spl, J.F. von Kurz)	?	?1752	HL	—	lost or = no.1b; 1st known perf. Vienna, 29 May 1753	7
1b	XXIXb:1b	Der neue krumme Teufel (Asmodeus, der krumme Teufel) (oc/Spl, ?, Kurz), incl. Arlequin, der neue Abgott Ram in Amerika (pantomimic Spl), Il vecchio ingannato (int)	?	c1759	lib	HW xxiv/2, 3 (lib)	music lost; int ? not by Haydn; 1770 with variant pantomimic Spl Die Insul der Wilden, HW xxiv/2, 23 (lib)	7
2	XXVIII:1	Acide (festa teatrale, 1, G.B. [?G.A.] Migliavacca, after P. Metastasio: Galatea)	2 S, A, T, B, 2 fl, 2 ob/ eng hn, 2 hn, str, bc	1762	A, EK	HW xxv/1, 1	frag., lib extant; perf. Eisenstadt, 11 Jan 1763; ov., cf Hla:5	14, 58

No.	H	Title, librettist	Forces	Date	Authentication	Edition	Remarks
		[2nd version]	2 S, T, 2 B with addl 2 bn	[1773/4]	A	HW xxv/1, 105	frag.; perf. Eszterháza, 25 Sept 1774
3	XXX:1	Marchese (La marchesa Nespola) (comedia)	2 S, T, 2 fl, 2 ob, 2 hn, str, bc	?1763	A, EK	HW xxv/1, 139	frag.; lib/dialogues lost 14, 58
4	XXIVb:1	[title unkown] (?ob)	S, B, ? other vv, 2 eng hn, 2 hn str, bc	?1761/2	A (u)	HW xxv/1, 201	? = no.5; aria Costretta a piangere and recit extant
5	ii, 448	Il dottore (comedia)	?	?c1761–5	EK	—	lost
6	ii, 448	La vedova (comedia)	?	?c1761–5	EK	—	lost
7	ii, 448	Il scanarello (comedia)	?	?c1761–5	EK	—	lost
8	XXVIII:2	La canterina (int in musica, 2)	2 S, T, 2 fl, 2 ob/eng hn, 2 hn, str, bc	1766	A, EK	HW xxv/2; facs. of lib, Haydn Yearbook 1996	perf. ?Eisenstadt, before 11 Sept (? 26 July) 1766; Bratislava, 16 Feb 1767; lib from int in Conforto: La commediante, 1754, and Piccinni: L'Origille, 1760, after Scioli: La cantarina, 1753; text of nos.2, 3 from A. Zeno: Lucio Vero, 1700; cf Q 29 17, 58
9	XXVIII:3	Lo speziale (Der Apotheker) (dg, 3, C. Goldoni, rev. ? C. Friberth)	2 S, 2 T, 2 fl, 2 ob, bn, 2 hn, str, bc	[1768]	A, EK	HW xxv/3; facs. of lib, Haydn Yearbook 1997	Act 3 inc.; perf. Eszterháza, aut. (? 28 Sept) 1768; ov., cf K 6 17, 58, 61
10	XXVIII:4	Le pescatrici (Die Fischerinnen) (dg, 3, Goldoni, rev. ??Friberth)	2 S, A, 2 T, 2 B, vv, 2 fl, 2 ob/eng hn, bn, 2 hn, str, bc	1759	A, EK	HW xxv/4; facs. of lib, Haydn Yearbook 1996	Acts 1, 2 inc.; 1st known perf. Eszterháza, 16, 18 Sept 1770; ?1st movt of ov., cf J 106 17, 59
11	XXVIII:5	L'infedeltà delusa (Liebe macht erfinderisch; Untreue lohnt sich nicht; Deceit outwitted) (burletta per musica, 2, M. Coltellini, rev. ?Friberth)	2 S, 2 T, B, 2 ob, 2 bn, 2 hn, timp, str, bc	[1773]	A, EK	HW xxv/5	perf. Eszterháza, 26 July 1773; ov., concert version, cf K 8 17, 59
12	XXIXa:1, 1a; XXIXb:2	Philemon und Baucis, oder Jupiters Reise auf die Erde (Spl/marionette op, 1, G.K. Pfeffel); Vorspiel: Der Götterrat (1, ? P.G. Bader)	2 S, 2 T, 4vv, ? 2 fl, 2 ob 2bn, 2 hn, ? 2 tpt, timp, str	[1773]	EK, signed lib, HL, A (frag., u)	HW xxiv/1	supposed ov. (cf J 50) and frag. of prelude extant; drama extant in rev. version; perf. Eszterháza, 2 Sept 1773; ov. to drama, HIa:8; cf appx G. 1, 1 5, 17, 59
13	XXIXa:2	Hexenschabbas (marionette op)	?	?1773	Dies	—	lost 17

No.	H	Title, librettist	Forces	Date	Authentication	Edition	Remarks
14	XXVIII:6	L'incontro improvviso (Die unverhoffte Zusammenkunft; Unverhofftes Begegnen) (dg, 3, Friberth, after Dancourt: La rencontre imprévue)	3 S, 2 T, 2 B, 2 ob/eng hn, 2 bn, 2 hn, 2 tpt, timp, perc, str, bc	[1775]	A, EK	HW xxv/6 (I, II)	perf. Eszterháza, 29 Aug 1775; ov., K 5	5, 17, 59
15	XXIXa:3	Dido (Spl/marionette op, 3, Bader)	?	[1775/6]	HL	HW xxiv/2, 31 (lib, 1778)	music lost; perf. Eszterháza, ? Feb/March 1776, also aut. 1778; ?aria extant (G 13)	20
16a	XXIXA:4	Opéra comique vom abgebrannten Haus	?	?c1773–9	EK	?	lost or = no.16b	20
16b	XXIXb:A	Die Feuersbrunst (Spl/?marionette op, 2)	S, ?5 T, B, 4vv, 2 fl, 2 ob, 2 cl, 2 bn, 2 hn, 2 tpt, timp, str	?1775–8	?	HW xxiv/3	? = no.16a; authenticity uncertain; dialogues lost; 1st, 2nd, ?3rd movts of ov. by I.J. Pleyel; cf K 8, appx K 1	20
17	XXVIII:7	Il mondo della luna (Die Welt auf dem Monde) (dg, 3, Goldoni)	2/3 S, 1/2 A, 2 T, B, 2 fl, 2 ob, 2 bn, 2 hn, 2 tpt, timp, str, bc	[1777]	A, EK, Sk	HW xxv/7 (I, II, III)	perf. Eszterháza, 3 Aug 1777; cf A 8, J 63, K 7, appx K 8, S 8, 9, 10, 12, 13, appx Y.3, 4	20, 22, 59, 61
18	XXIXb:3	Die bestrafte Rachbegierde (Spl/marionette op, 3, Bader)	?	?1779	lib	HW xxiv/2, 57 (lib)	music lost; perf. Eszterháza, 1779	20
19	XXVIII:8	La vera costanza (dg, 3, F. Puttini)	?	–1779 [? April–Nov 1778]	EK, Sk	—	music lost where nor incl. in 2nd version; sketches, HW xxv/8, 356; perf. Eszterháza, 25 April 1779; ov., concert version, K 7	20, 59–60
		2nd version (Der flatterhafte Liebhaber; Der Sieg der Beständigkeit; Die wahre Beständigkeit; List und Liebe; Laurette (P.U. Dubuisson) (nXXVIII: 8a))	3 S, 3 T, B, 1/2 fl, 2 ob, 2 bn, 2 hn, timp, str, bc	1785	A (partly in copyists' hands)	HW xxv/8	Count Errico's Act 2 scene = that in Anfossi's setting (1775)	
20	XXVIII:9	L'isola disabitata (Die wüste Insel) (azione teatrale, 2, Metastasio)	2 S, T, fl, 2 ob, bn, 2 hn, timp, str, bc	1779	A (frags.), EK	(Vienna and Leipzig, 1909) (vs)	perf. Eszterháza, 6 Dec 1779, finale rev. 1802; ov., K 4, autograph frag. in PL-Kj	6, 20, 60
21	XXVIII:10	La fedeltà premiata (Die belohnte Treue) (dramma pastorale giocoso, 3, after G. Lorenzi: L'infedeltà fedele)	4 S, 2 T, 2 B, fl, 2 ob, bn, 2 hn/tpt, timp, str, bc	1780	A, RC	HW xxv/10 (I, II)	perf. Eszterháza, 25 Feb 1781; cf appx E 3, J 73	20, 22, 46, 60
22	XXVIII:11	Orlando paladino (Der Ritter Roland) (dramma eroicomico, 3, C.F. Badini, N. Porta)	3 S, 4 T, 2 B, fl, 2 ob, 2 bn, 2 hn/tpt, timp, str, bc	1782	A	HW xxv/11 (I, II)	perf. Eszterháza, 6 Dec 1782; ov., vla.16; cf D 7; duetto H 16 arr. with text Quel cor umano e tenero (L. da Ponte) (London, 1794–5)	19, 20, 60

No.	H	Title, librettist	Forces	Date	Authentication	Edition	Remarks	
23	XXVIII:12	Armida (dramma eroico, 3, after I. Durandi, F.S. De Rogati and others)	2 S, 3 T, B, fl, 2 ob, 2 cl, 2 bn, 2 hn/tpt, timp, str, bc	1783	A	HW xxv/12	perf. Eszterháza, 26 Feb 1784; ov, HIa:14	20, 60
24	XXVIII:13	L'anima del filosofo, ossia Orfeo ed Euridice (dramma per musica, 4/5, Badini)	2 S, T, B, 4vv, 2 fl, 2 ob, 2 cl, 2 eng hn, 2 bn, 2 hn, 2 tpt, 2 trt-n, timp, hp, str, bc	1791	A	HW xxv/13	perf. Florence, 9 June 1951; ov, HIa:3, cf K 13; chorus Finché circola il vigore uses final chorus from Orlando paladino (E 22); cf D 7	32, 33, 60, 61
25	XXX:5	Alfred, König der Angelsachsen, oder Der patriotische König (J.W. Cowmeadow, after A. Bicknell):		1796	A		incid music; perf. as Haldane, König der Dänen, Eisenstadt, 9 Sept 1796	
a		Triumph dir, Haldane (chorus)	3vv, 2 ob, 2 bn, 2 tpt, timp, str			(Leipzig, 184) (vs)		
b		Ausgesandt vom Strahlenthrone (aria with spoken interjections)	S, 2 cl, 2 bn, 2 hn			(Salzburg, 1961)		
c		Der Morgen graut (duet)	2 T, ?hp, vn solo str					
26	XXX:4	Fatal amour (recit and aria with spoken interjections), F, G, Eb	S, fl, 2 ob, 2 bn, 2 hn, str	?c1796	A (u)	—		

Note: an 'operette' mentioned by Elssler in 1811, not identified (see Schmieder, *AMz*, lxv (1927), 425–7)

Appendix E: Selected doubtful and spurious works

No.	H	Title, librettist	Edition	Remarks
1	XXIXa:5	Genovefens vierter Theil (Spl/marionette op, 3, K. von Pauersbach)	HW xxiv/2, 75 (lib)	music lost; by different composers according to HL, by Haydn according to HV; perf. Eszterháza, sum. (? 6 Aug) 1777; ov, ? K 3
2	XXIXb:F Add.	Die reisende Ceres (Spl, M. Lindemay?)	(Vienna, 1977)	music inc.
3	XXXII:2	Der Freibrief (Spl, 1, ? G.E. Lüderwald)	—	music lost; pasticcio incl. music from La fedeltà premiata (E 21), ? arr. F. vor. Weber, perf. Meiningen, 1789
4	XXXII:3	Alessandro il grande (3s, 3)	—	pasticcio arr. J. Schellinger from works by Haydn and others, dated 1792(?0)
5	XXXII:4	Der Äpfeldieb (Spl, 1, C.F. Bretzner)	—	music lost; perf. Hamburg, 1791, with inserted music by Haydn

No.	H	Title, librettist	Edition	Remarks
6	i, 577	Die [Das] Ochsenmenuett (Singspiel, 1, G.E. von Hofmann)	(Mainz, 1927)	pasticcio arr. I. von Seyfried from Haydn's works; perf. Vienna, 1823; see appx X.3, 8
7	—	Das Teebrett (comedy, E. Fischer), vv, kbd	(Berlin, 1914)	music from L'incontro improvviso (E 14), L'infedeltà delusa (E 11), Orlando paladino (E 22)
8	—	(Finale ... sey voll edlen Stolzes)	—	2 coloraturas, S, orch; without author's name; MS (D-LEm) not a Haydn autograph

F: SECULAR VOCAL WITH ORCHESTRA

No.	H	Title	Forces	Date	Authentication	Edition	Remarks	
1	XXIVb:A1	Aure dolci ch'io respiro (aria)	?S/T, 2 fl, 2 ob, str	−?1762	F	—	vocal part lost	
2	XXIVb:2	D'una sposa meschinella	S, 2 ob, 2 hn, str	?sum. 1777	A	(Salzburg, 1961)	aria for Paisiello: La Frascatana; ? by unknown composer, ?rev. Haydn; ? orig. = hXXIVb:2bis	27, 61
3	XXIVb:8	Dica pure chi vuol dire	?S, 2 ob, bn, 2 hn, str	?1778/85	—	(Vienna, 1787) (vs)	aria for Anfossi: Il geloso in cimento; newly scored P.A. Pisk (Vienna, 1931)	
4	XXIVb:3	Quando la rosa ... Finché l'agnello	S, fl, bn, 2 hn, str	?July 1779	A (u)	(Salzburg, 1961) stanza only	(1st aria for Anfossi: La Metilde ritrovata (L'incognita perseguitata); recit is by Anfossi	
5	XXIVb:5	Dice benissimo	B, 2 hn, str	−?27 July 1780	A (frags., u)	(Salzburg, 1964)	aria for Salieri: La scuola de' gelosi; also with texts Männer ich sag es euch and Ja in dem Himmel	
6	XXIVb:7	Signor voi sapete	S, 2 fl, 2 ob, 2 bn, 2 hn, str	−23 July 1785	Sk, HE	(Salzburg, 1961)	aria for Anfossi: Il matrimonio per inganno; sketch in PL-Kj	
7	XXIVa:7	Miseri noi ... Funesto orror	S, 2 fl, 2 ob, 2 bn, 2 hn, str	−1786	SC	D 17 (1960)	cant. (recit and aria)	
8	XXIVb:9	Sono Alcina	S, fl, 2 ob, 2 bn, 2 hn, str	[−18 June] 1786	A	(Salzburg, 1961)	cavatina for G. Gazzaniga: L'isola di Alcina; cf T 3	
9	XXIVb:10	Ah tu non senti ... Qual destra omicida	T, fl, 2 ob, 2 bn, 2 hn, str	[−4 July] 1786	A, Sk	(Salzburg, 1964)	recit and aria for Traetta: Ifigenia in Tauride; sketch ed. A.P. Brown (K1979)	
10	XXIVb:11	Un cor si tenero	B, 2 ob, 2 hn, str	[−April] 1787	A	(Salzburg, 1964)	aria for F. Bianchi: Il disertore	
11	XXIVb:12	Vada adagio, signorina	S, 2 ob, 2 bn, 2 hn, str	−23 June 1787	A (u pt), C	(Salzburg, 1961)	aria for P. Guglielmi: La Quakera spiritosa: cf appx F.2, 3	
12	XXIVb:13	Chi vive amante	S, fl, 2 ob, 2 bn, 2 hn, str	[−25/26 July] 1787	A	(Salzburg, 1961)	aria for F. Bianchi: Alessandro nell'Indie	
13	XXIVb:14	Se tu mi sprezzi	T, 2 ob, 2 bn, 2 hn, str	[−9 March] 1788	A	(Salzburg, 1964)	aria for G. Sarti: I finti eredi	
14	XXIVb:15	Infelice sventurata	S, 2 ob, 2 bn, 2 hn, str	[−Feb] 1789	A	(Salzburg, 1961)	aria for Cimarosa: I due supposti conti	
15	XXXII:1	for Circe, ossia L'isola incantata:		[−July] 1789			pasticcio by Anfossi; La maga Circe, Naumann and Haydn, cf appx F.2, 19; Lavatevi presto, cf appx X.3, 2	

Joseph Haydn: portrait (c1768) by
Johann Grundmann, probably destroyed
in World War II

Joseph Haydn: anonymous miniature
(?c1788) in the Gesellschaft der
Musikfreunde, Vienna

Autograph MS of part of Haydn's 'Salve regina' in E,
composed 1756 (D-DS 988, f.2v)

A page from Haydn's 'Entwurf-Katalog', begun in 1765 (*D-Bsb*)

Joseph Haydn: engraving (c1792) by Luigi Schiavonetti after the second version of a portrait by Ludwig Guttenbrunn, the first version of which probably dates from c1770

Advertisement from 'The Times' (16 May 1791) concerning
Haydn's benefit concert, to be given on that day

Joseph Haydn: bust (1801–2)
by Anton Grassi in the Historisches
Museum der Stadt Wien

Joseph Haydn: portrait (1791)
by John Hoppner, Royal Collection,
Windsor Castle

Joseph Haydn: pencil portrait
(1794) by George Dance in
the Historisches Museum der
Stadt Wien

Autograph MS from Haydn's String Quartet in C
op.20 no.2, composed 1772 (*A-Wgm*)

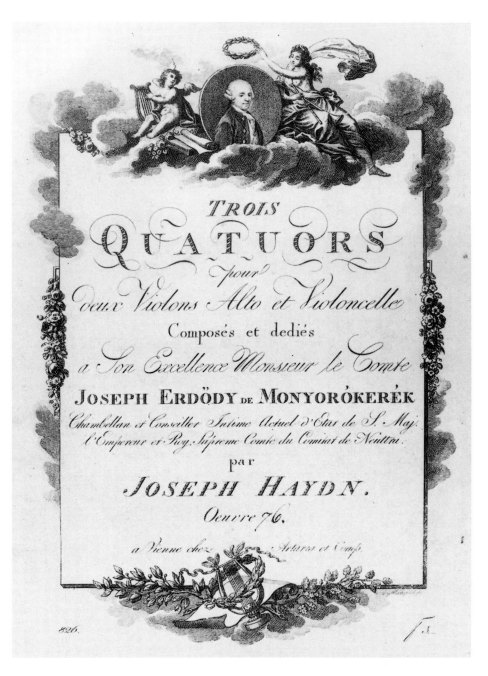

Title-page of the 'Erdődy' Quartets op.76, published by Artaria in 1799

Autograph sketches for Haydn's 'The Seasons', composed 1799–1801 (D-Bsb autogr. Jos. Haydn 50, f.2v)

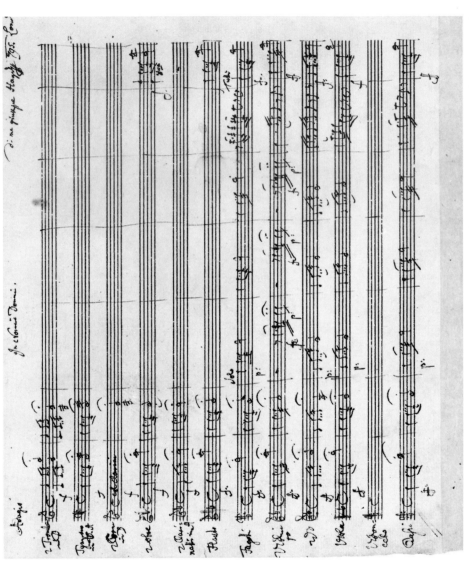

Opening page of the autograph MS of Haydn's Symphony no.104 in D ('London'), composed 1795 (*D-Bsb*)

No.	H	Title	Forces	Date	Authentication	Edition	Remarks
	a	Son due ore che giro (recit)	T, fl, 2 ob, 2 bn, str, bc		A	(Budapest, 1960) (see Bartha and Somfai, E1960)	
	b	Son pietosa, son bonina (aria)	S, fl, 2 ob, 2 bn, 2 hr, str		C	D 19 (1959)	
	c	Lavatevi presto (terzetto)	2 T, B, fl, 2 ob, 2 bn, 2 hn, str		C	(Cardiff, 1982)	
16	XXIVb:16	Da che penso a maritarmi	T, fl, 2 ob, 2 bn, 2 hr, str	[-14 March] 1790	A	(Salzburg, 1964)	aria for Gassmann: L'amore artigiano, which incl. 2 other arias by Haydn: ?no.17, ?appx F1, 5
17	XXIVb:19	La mia pace, oh Dio, perdei (aria)	S, fl, 2 ob, 2 bn, 2 hn, str	1790	A	—	see no.16
18	XXIVb:17	Il meglio mio carattere	S, fl, 2 ob, 2 bn, 2 hn, str	-26 June 1790	C	(Salzburg, 1961)	aria for Cimarosa: L'impresario in angustie
19	XXIVb:18	La moglie quando è buona	S, fl, 2 ob, 2 bn, 2 hn, str	-?Aug/Sept 1790	HE, C	(Salzburg, 1961)	aria for Cimarosa: Giannina e Bernardone; cf appx H 13
20	XXIVb:22*	Tornate pur mia bella (aria)	T, fl, 2 ob, 2 bn, 2 hn, str	-13 Aug 1790 [?1787]	—	—	with text Consola pur mia bella inserted in Guglielmi: La Quakera spiritosa, Vienna, 1790
21	XXIVb:23*	Via siate bonino (aria)	S, fl, 2 ob, 2 br, 2 hn, str	?-1785–95	—	—	
22	XXIVb:24	Cara deh torna, aria for (Giacomo) Davide	T, ob, bn (and ?)	-16 May 1791	Gr, Dies	—	music lost; text in Landon (A1976), 76
23	—	Aria for Miss Poole	?S, ?	1791–5	Gr, Dies	—	lost/unidentified; ?sketch (see Feder, E1980)
24	—	(aria with full orch)	?	1791–5	Gr, Dies	—	lost/unidentified
25	XXIVa:10	Berenice, che fai (cant.)	S, fl, 2 ob, 2 cl, 2 bn, 2 hn, str	[-4 May] 1795	A	D 129 (1965)	scena from Metastasio: Antigono; 36, 62 composed in London for Brigida Giorgi Banti
26	XXIVb:20	Solo e pensoso (aria)	S, 2 fl, 2 bn, 2 hn, str	1798	A	(Salzburg, 1961)	soretto no.28 from Petrarch: Canzoniere (r.o.xxxv)

Note. Ah come il core (HXXIVa, Anh.4), see E 21 (HW xxv/10, 380); Costretta a piangere, see E 4; Quel cor umano e tenero (HXXIVa, Anh.) = Quel tuo visetto amabile, see E 22 (HW xxv/11, 237); Sono le donne capricciose = Dice benissimo (no.5)

Frag., E(Eb), MGG1, v, 1893, line 4 = Dice benissimo (no.5)

Unpubd secco recits, rev./composed Haydn, in A. Felici (?Sacchini): L'amore soldato, perf. Esterháza, 1779, and Cimarosa: L'impresario in angustie, perf. Esterháza, 1790

Recit and aria sung by Calcagni, London, 1792, and Cantata a voce sola con violino composed for the Duke of Bedford, unidentified or lost; see Landon (A1976)

Appendix F.1: Selected works attributed to Haydn

No.	H	Title	Forces	Date	Edition	Remarks
1	XXVb:5*	Pietà di me, benigni Dei (terzetto)	2 S, T, eng hn solo, bn solo, hn solo, 2 hn, str	?	D 250 (1982)	Haydn's name on MS copies added later; considered as probably authentic by Landon (A1978, A1976) and by Larsen (B1941)
2	XXIVb:6	Mora l'infido … Mi sento nel seno (recit and aria)	S, orch	[1781]	—	extant are vocal part and 2 ob of recit, 2 vn (partly in Haydn's hand) and b of recit and aria; without author's name; incl. in Righini: Il convitato di pietra
3	XXVIb:1	Er ist nicht mehr! (Deutschlands Klage auf den Tod des grossen Friedrichs, Borussens König) (cant.)	?S, baryton (and ?)	1786–8	—	perf. Leipzig and Nuremberg, 1788, by Carl Franz; vocal part with bc extant; considered authentic by Landon (A1978)
4	XXIVb:21*	(aria)	?S, 2 fl, 2 ob, 2 bn, 2 hn, str	–1788	—	vocal part lost; without author's name; incl. in G. Sarti: I finti eredi; cf appxs X.1, 7, Y.4, 1
5	XXIVb:16bis	?Occhietti cari del mio tesoro (aria)	?T, 2 ob, bn, 2 hn, str	–1790	—	vocal part lost; without author's name; incl. in Gassmann: L'amore artigiano; see F 16
6	XXIVa:F1	Pianger vidi appresso un fonte (cant.)	A, 2 ob, 2 hn, str	?		probably not authentic
7	XXIIId:B2	Veni tandem expectatus (aria)	S, str	?	(Munich, 1942)	not sacred; without author's name; not a Haydn autograph

Note: several anon. 'Teutsche Comoedie-Arien' (arias from Viennese Singspiele of 1750s) tentatively attrib. Haydn; 22 It. arias from Esterházy archives listed in Bartha and Somfai (E1960) as probably by Haydn, though no source with his name is known; more anon. arias from same archives mentioned in Landon (A1978) as possibly by Haydn

Appendix F.2: Revisions (mostly in Haydn's hand) of operatic works by other composers

No.	H	Title	Forces	Date	Edition	Remarks
1	XXXIc:3	Vi miro fiso	?S, 2 ob, 2 hn, str	[aut. 1777]	—	aria from Dittersdorf: L'Arcifanfano re de' matti; altered and wind pts added
2	—	Non per parlar d'amore	?S, orch	[July 1778]	—	aria by Salieri from pasticcio L'astratto; altered and 8 bars rewritten; see Bartha and Somfai (E1960), iii/2
3	XXXIc:4	?Se provasse un pocolino	?S, ? 2 fl, 2 ob, 2 hn, str	[Feb 1780]	—	aria from Anfossi: La forza delle donne, inc.; wind pts added and 2 vn rewritten; melody similar to F 11
4	XXXIc:5	Ah crudel, poiché lo brami	S, 2 fl, 2 hn, str	[April 1780]	(Salzburg, 1961)	aria from G. Gazzaniga: La vendemmia; 2nd half composed by Haydn
5	XXIVb:4	?	S, 2 ob, 2 hn, str	[July] 1780]	—	aria incl. in Salieri: La scuola de' gelosi; text lost, not Il cor nel seno; without author's name; score mostly rewritten and perhaps composed by Haydn
6	XXXIc:6	Gelosia d'amore è figlia (2 versions)	S, 2 ob, 2 hn, str	[July 1780]	—	aria from Salieri; ibid; altered, wind pts added, 34 bars of score added or rewritten; sketch in HW xxix/2, 82

No.	H	Title		Date	Edition	Remarks
7	XXXIc:7	Si promette facilmente	S, 2 ob, 2 hn, str	[Oct 1780]	—	aria from Anfossi: La finta giardiniera; completely rewritten
8	XXXIc:8	Vorrei punirti indegno	S, 2 ob, 2 hn, str	[Oct 1780]	—	aria from Anfossi, ibid.; altered and wind pts added
9	XXXIc:9	Non ama la vita	?S, orch	[April 1781]	—	aria from Anfossi: Isabella e Rodrigo, ossia La costanza in amore; 2 bn and 6 bars added
10	XXXIc:10	Che tortora	S, 2 ob, 2 hn, str	[Aug 1781]	—	aria from N. Piccinni: Gli stravaganti, ossia La schiava riconosciuta; 70 bars rewritten
11	XXXIc:11	Una semplice agnelletta	S, orch	[Aug 1781]	—	aria from Piccinni, ibid.; altered, 14 bars added
12	—	Siam di cuor tenero	S, str (and ?)	[Aug 1781]	—	aria from Piccinni, ibid.; 8 (?7 + 3) bars rewritten; see Bartha and Somfai (E1960), iii/14
13	—	Misera che farò	S, orch	[March 1782]	—	recit from Traetta: Il cavaliere errante; only 2 vn extant; not autograph; authenticity uncertain; see Bartha and Somfai (E1960), iii/23
14	XXXIc:12	Deh frenate i mesti accenti	?S, 2 ob, 2 hn, str	[Sept 1782]	—	aria from Anfossi: Il curioso indiscreto; extensively rev., wind pts added, 2 fl, bn omitted
15	—	Dove mai s'èritrovato	S, orch	[March 1784]	—	aria from Anfossi: I viaggiatori felici; 6 bars rewritten; see Bartha and Somfai (E1960), iii/16
16	—	Ah mi palesa almeno	S, T, orch	[July 1786]	—	duet from Traetta: Ifigenia in Tauride; vocal parts and 2 bars of score rewritten; see Bartha and Somfai (E1960), iii/17
17	XXXIc:15	Se palpitar degg'io	S, ob, str (and ?)	[Aug 1788]	—	aria from Prati: La vendetta di Nino (Semiramide); 27 bars added
18	XXXIc:13	Se voi foste un cavaliere	S, str	[Feb 1789]	—	aria from Cimarosa: I due supposti conti; completely rewritten
19	—	Quasi in tutte le ragazze	S, orch	[July 1789]	—	aria in pasticcio Circe (F 15); 12 bars rewritten; see Bartha and Somfai (E1960), iii/20
20	XXXIc:14	Silenzio, miei signori	T/B, orch	[June 1790]	—	from quintet in Cimarosa: L'impresario in angustie; 25 bars added

Note: HXXXIc:2 shows only minor alterations, as do many other arias described in Bartha and Somfai; for added or altered parts not written by Haydn, and the revision therefore of doubtful authorship, see Bartha and Somfai (E1960), iii/8*, 10*, 11*, 18*

89

G: SOLO SONGS WITH KEYBOARD

No.	HXXVIa	Title, poet, key	Date	Authentication	Edition	Remarks	
1–36	1–12	XII Lieder für das Clavier, i:	–27 May 1781	OE, ?EK, HC	HW xxix/1, 2–16		25, 62
		1 Das strickende Mädchen (Sir Charles Sedley, trans. J.G. Herder), Bb; 2 Cupido (G. Leon), E; 3 Der erste Kuss (J.G. Jacobi), Eb; 4 Eine sehr gewöhnliche Geschichte (C.F. Weisse), G; 5 Die Verlassene (L.L. Haschka), g; 6 Der Gleichsinn (G. Wither, trans. J.J. Eschenburg), A; .7 An Iris (J.A. Weppen), Bb; 8 An Thyrsis (1st stanza: C.M. von Ziegler, rest anon.), D; 9 Trost unglücklicher Liebe, f; 10 Die Landlust (Stahl), C; 11 Liebeslied (Leon), D; 12 Die zu späte Ankunft der Mutter (Weisse), Eb					
	13–24	XII Lieder für das Clavier, ii:	1781 [?1780] –[? 3 March] 1784	OE, EK (nos.17, 24), A (no.18, u), Sk (no.19)	HW xxix/1, 17–31	sketch of no.19 in HW xxix/2, 82	25, 62
		13 Jeder meint, der Gegenstand (P.G. Bader), F (? from Dido (E 15), cf appx Y.3, 1); 14 Lachet nicht, Mädchen, Bb; 15 O liebes Mädchen, höre mich, G; 16 Gegenliebe (G.A. Bürger), G (cf J 73); 17 Geistliches Lied, g; 18 Auch die sprödeste der Schönen (F.W. Gotter), F; 19 O fliess, ja wallend fliess, E; 20 Zufriedenheit (J.W.L. Gleim), C; 21 Das Leben ist ein Traum (Gleim), Eb; 22 Lob der Faulheit (G.E. Lessing), a; 23 Minna (J.J. Engel), A; 24 Auf meines Vaters Grab, E					22
	25–30	VI Original Canzonettas (A. Hunter), i:	–3 June 1794	OE, Sk (nos.29, 30)	HW xxix/1, 34–51	sketches of nos.29, 30 in HW xxix/2, 83	36, 62
		25 The Mermaid's Song, C; 26 Recollection, F; 27 Pastoral Song, A; 28 Despair, E; 29 Pleasing Pain, f [nos.29, 30: –?19 Jan 1794]; 30 Fidelity, f (cf B 22)					
	31–6	VI Original Canzonettas, ii:	–14 Oct 1795	RC (no.31), Sk (no.32), HE (nos. 33, 34), EK [no.32: –?19 Jan 1794]	HW xxix/1, 52–69	sketch of no.32 in HW xxix/2, 86 text of no.34 from Shakespeare: *Twelfth Night*; no.36 with 2 texts, see critical commentary to HW xxix/1 and facs. (Cardiff, 1983)	62
		31 Sailor's Song, A; 32 The Wanderer (Hunter), g; 33 Sympathy (J. Hoole, after Metastasio: *L'olimpiade*), E; 34 She never told her love (W. Shakespeare), Ab; 35 Piercing Eyes, G; 36 Transport of Pleasure [Content], A					
36b		Der verdienstvolle Sylvius (Ich bin der Verliebteste) (J.N. Görz), Ab	–1 Feb 1795 [–?1788]	Sk, HE, Gr	HW xxix/1, 70	orig. version of no.36; sketch in HW xxix/2, 88	
37–47	37	Beim Schmerz, der dieses Herz durchwühlet, E	?c1765–75	A (u), HC, HV	HW xxix/1, 74	? part of dramatic work	
	38	Der schlau(e) und diensfertige Pudel (v.T. . . .), Bb	c1780–87	A (u), Gr, Dies	HW xxix/1, 76		
	39	Trachten will ich nicht auf Erden, E	–14 Dec 1790	A	HW xxix/1, 78	date on autograph is that of ded.; facs. see critical commentary to HW xxix/1, 16	
	40	Der Feldzug	?	HC	—	lost or unidentified	
	41	The Spirit's Song (Hunter), f	–9 Sept 1800 [?c1795]	E, HV	HW xxix/1, 81		
	42	O Tuneful Voice (Hunter), Eb	?c1795	Gr	HW xxix/1, 84		62
	43	Gott, erhalte [Franz] den Kaiser! (Haschka), G	Oct 1796–Jan 1797	A, Sk	HW xxix/1, 89	facs. often publd; used as Ger. and former Austrian national anthem; sketch in HW xxix/2, 90; cf D 11, O 62	62

No.	HXXVIa	Title, poet, key	Date	Authentication	Edition	Remarks
	44	Als einst mit Weibes Schönheit, A	?c1796–1800	A (u)	HW xxix/1, 90	
	45	Ein kleines Haus, E	–30 Aug 1800 [–?1797]	A	HW xxix/1, 92	autograph signed later, 20 July 1807 (?1801); facs. see Sandberger, ZfM, cix (1942), 535–8
	46	Antwort auf die Frage eines Mädchens, G	–June 1803	SC	HW xxix/1, 95	signed MS in PL-Kj; title Vergiss mein nicht not authentic
	47	Bald wehen uns des Frühlings Lüfte, G	?	E (without authors name)	HW xxix/1, 98	2nd stanza lost
48–51	48a–d	Four German Songs:	?	A (incipits only, u)	HW xxix/1, 99 (incipits)	lost; ? popular tunes arr. Haydn; for text of no.50 see P. Dormann: *Franz Joseph Aumann* (Munich, 1985), 410, 414; arr. of no.51 by A. Albrechtsberger with text Kein lustiges Leben in A-Wgm; cf P 7
		48 Ich liebe, du liebest, E♭; 49 Dürre, Staub, B♭, 50 Sag 'n allweil (? M. Lindemayr), C; 51 Kein besseres Leben, G				

Note: further songs, mentioned by Griesinger and Dies as composed in England, may be identical with some of those listed above; 7 Ger. songs mentioned by Rosenbaum as perf. 16 Oct 1799 = ? some of H 6–18

Appendix G.1: Arrangements

No.	H	Title	Date	Authentication	Edition	Remarks
1	ii, 443	Canzonetta: Ein Tag, der allen Freude bringt (G.K. Pfeffel), A	?1773	A (u)	HW xxiv/1, 98	arr. from aria in E 12
2	XXXIc 17	The Lady's Looking-glass, D	?1791–5	A (u)	HW xxix/1, 97	arr. from catch, 3vv, by Earl of Abingdon; followed by short kbd piece, D (X 7); cf S 15

Appendix G.2: Selected spurious works

No.	HXXVIa	Title, poet, key	Edition	Remarks
1	F1	Abschiedslied, F	HW xxix/1, 79	by Gyrowetz
2	D4	Hymne an die Freundschaft, G	M 20/I, 111	arr. Küttner, based on II of J 75
3	D1	Liebes Mädchen, hör mir zu, D	M 20/I, 110	also known as Ständchen, 3vv (HXXVb:G1); also attrib. Mozart (K Anh.C9.04)
4	C1	Die Teilung der Erde (F. von Schiller), C	M 20/I, 112	by F. Roser von Reiter

No.	HXXVIa	Title, poet, key	Edition	Remarks
5	G1	A Prey to Tender Anguish (Ich habe viel gelitten), G	(London, 1797)	
6	Es4	Heiss mich nicht reden (J.W. von Goethe), Eb	(Vienna, 1925)	by Zumsteeg

H: MISCELLANEOUS VOCAL WORKS WITH KEYBOARD

No.	H	Title, poet, key	Forces	Date	Authentication	Edition	Remarks	
1	XXVIb:2	Arianna a Naxos (Teseo mio ben) (cant.)	S, hpd/pf	–9 Feb 1790	A (lost), OE	HW xxix/2, 2		29, 32, 62
2a	—	Maccone (Gesänge) for Gallini	?	1791–5	Gr, Dies	—	lost	
2b	—	Italian catch	?7vv, (?bc)	–2 June 1791	see Landon (A1976)	—	lost, ? partly = no.2a	
2c	—	Salomon und David	?	–1795	Haydn's 3rd London notebook	—	lost	
3	XXVIb:3	Dr. Harington's Compliment (What art expresses; Der Tausenden), A	S, 4vv, pf	?2–6 Aug 1794	Gr	HW xxix/2, 58	variations on song by Dr H. Harington	
4–5		2 Duetti of Nisa and Tirsi (C.F. Badini)	S, T, hpd	1796		D35 (1960)		
	XXVa:2	Saper vorrei, G			RC	HW xxix/2, 24		
	XXVa:1	Guarda/Senti qui, F			A	HW xxix/2, 34		
6–18		Aus des Ramlers Lyrischer Blumenlese (13 partsongs):	3–4vv, bc (nos.1–9)/ hpd obbl (nos. 10–13)	1796 (–?1799)	A	HW xxx	mentioned in letter to E.L. Gerber, 23 Sept 1799; Haydn used 1st bars of no.5 for his visiting-card in 1806; pf obbl for nos.1–9 added ? A.E. Müller	41, 62
	XXVc:1	1–Der Augenblick (J.N. Götz), A	S, A, T, B, bc					
	XXVc:2	2–Die Harmonie in der Ehe (Götz), Bb	S, A, T, B, bc					
	XXVc:3	3–Alles hat seine Zeit (Athenaeus, trans. J.A. Ebert), F	S, A, T, B, bc					
	XXVc:4	4–Die Beredsamkeit (G.E. Lessing), Bb	S, A, T, B, bc					
	XXVc:5	5–Der Greis (J.W.L. Gleim), A	S, A, T, B, bc					
	XXVb:1	6–An den Vetter (C.F. Weisse), G	S, A, T, bc					
	XXVb:2	7–Daphnens einziger Fehler (Götz), C	T, T, B, bc					
	XXVc:6	8–Die Warnung (Athenaeus, trans. Ebert), Bb	S, A, T, B, bc					

No.	H	Title, poet, key	Forces	Date	Authentication	Edition	Remarks
	XXVb:3	9–Betrachtung des Todes (C.F. Gellert), a	S, T, B, bc				
	XXVc:7	10–Wider den übermut (Gellert), A	S, A, T, B, hpd				
	XXVb:4	11–An die Frauen (Anakreon, trans. G.A. Bürger), F	T, T, B, hpd				
	XXVc:8	12–Danklied zu Gott (Gellert), Eb	S, A, T, B, hp d				
	XXVc:9	13–Abendlied zu Gott (Gellert), E	S, A, T, B, hpd				
19	XXVIb:4	The Battle of the Nile (Ausania trembling ... Blest leader) (Pindarick Ode) (cant., E.C. Knight)	1v, hpd/pf	?5–9 Sept 1800	RC (partly A)	HW xxix/2, 42	10 of 17 stanzas set
20–25	ii, 533	6 airs with variations (6 Admired Scotch Airs): 1 The blue bell(s) of Scotland (? Mrs Grant), D; 2 My love she's but a lassie yet (? H. Macneill), C; 3 Bannocks o'barley meal (? A. Boswell), G; 4 Saw ye my father? (? R. Burns), D; 5 Maggy Lauder, A; 6 Killicrankie (? Mrs Grant; ?Burns), C	1v, vn, vc, pf	1801/2–?	E (nos.2–6, without text, orig. without author's name); no.1: A (without text; vn, vc missing) and E (vn, vc only)	(London, 1805). arr. vn, pf (?fl)	each with 3 variations; themes (? and texts) = or nearly = Z 37, 242, 15, 296, 208, 175; date on autograph, 6 Feb 1805 [not 1806], is that of ded.

Note: Cant., 1v, vn (and ?), composed for Duke of Bedford, mentioned in Landon (A1976), lost or unidentified

Appendix H: Arrangements

No.	H	Title, key	Forces	Date	Authentication	Edition	Remarks
1–12	XXXIc:16	12 Sentimental Catches and Glees: 1 Know then this truth, A; 2 O say what is, G; 3 Hail to the myrtle shade, A; 4 Love free as air, D; 5 Ah no[n] lasciarmi, C; 6 O ever beauteous, A; 7 Where shall a hapless, G; 8 Ye little loves, Eb; 9 Some kind angel, A; 10 I fruitless mourn, a; 11 Farewell my flocks, A; 12 The envious snow, C	3vv, hp/pf	1795	?Gr, ?Dies	HW xxix/2, 66	melodies by Earl of Abingdon, acc. (? and 3vv settings) by Haydn; nos.3, 7, 8 glees, others catches
13	ii, 217	La moglie quando è buona, aria, Eb	S, hpd	?1790–98	C	—	arr. of F 19

I: CANONS

HXXVII	Title, poet, key, forces	Date	Authentication Edition
a: 1–10	Die Heiligen Zehn Gebote als Canons (The Ten Commandments): 1 Canon cancrizans: Du sollst an einen Gott glauben, C, 3/4vv; 2 Du sollst den Namen Gottes nicht eitel nennen, G, 4vv; 3 Du sollst Sonn- und Feiertag heiligen, B♭, 4vv; 4 Du sollst Vater und Mutter verehren, E♭, 4vv; 5, 5b Du sollst nicht töten, g, 4vv (2 versions); 6 Du sollst nicht Unkeuschheit treiben, C, 5vv; 7 Du sollst nicht stehlen, a, 5vv; 8 Du sollst kein falsch Zeugnis geben, E, 4vv; 9 Du sollst nicht begehren deines Nächsten Weib, C, 4vv; 10 Du sollst nicht begehren deines Nächsten Gut, f, 4vv	c1791–5	A, A (u, no.5 b), Sk (nos.1, 5b, 7) — HW xxxi, 3–18; cf critical commentary, 8ff; for 5b (not in H), see Haydn-Studien, iv/1 (1976), 53
b: 1–47	40 (recte: 46/47) Sinngedichte als Canons bearbeitet: 1 Hilf an Narziss (F. von Hagedorn), g, 3vv; 2 Auf einen adeligen Dummkopf (G.E. Lessing), E♭, 3vv; 3 Der Schuster bleib bei seinem Leist (Das Sprichwort; Carone in carricatura) (K. von Eckartshausen), F, 8vv; 4 Herr von Gänsewitz zu seinem Kammerdiener (G.A. Bürger), c, 4vv; 5 An den Marull (Lessing), F, 5vv; 6 Die Mutter in ihr Kind in der Wiege, E♭, 3vv (4th v added M. Haydn, cf HW, critical commentary, 23); 7 Der Menschenfreund (Gellert), E♭, 4vv; 8 Gottes Macht und Vorsehung (Gellert), G, 3vv; 9 An Dorilis (K.F. Kretschmann), F, 4vv; 10 Vixi (Horace), B♭, 3vv; 11 Der Kobold (M.G. Lichtwer), E♭, 4vv; 12 Der Fuchs und der Marder (Lichtwer), a, 4vv; 13 Abschied, B♭, 5vv; 14 Die Hofstellungen (F. von Logau), b, 3vv; 15 Aus Nichts wird Nichts (Nichts gewonnen, nichts verloren) (A. Blumauer, after M. Richey), C, 5vv 16 Cacatum non est pictum (Bürger), A, 4vv; 17 Tre cose (G.A. Federico), E♭, 3vv; 18 Vergebliches Glück (trans. from Arabic A. Tscherning), A, 2vv; 19 Grabschrift (P.W. Hensler), g, 4vv (? originally planned as partsong); 20 Das Reitpferd (Lichtwer), E♭, 3vv; 21 Tod und Schlaf (Logau), f, 4vv; 22 An einen Geizigen (Lessing), D, 3vv; 23, 23b Das böse Weib (Lessing), g, 3vv; ?C, 2vv (2 versions); 24 Der Verlust (Lessing), E, 3vv; 25 Der Freigeist, G, 3vv; 26 Die Liebe der Feinde (Gellert), A, 2vv; 27 Der Furchtsame (Lessing), E♭, 4vv; 28 Die Gewissheit (Lessing), E♭, 4vv; 29 Phöbus und sein Sohn (Lichtwer), G, 4vv 30 Die Tulipane (Lichtwer), ?C, 2vv; 31 Das grösste Gut, ?C, 2/3vv; 32 Der Hirsch (Lichtwer), d, 5vv; 33 Überschrift eines Weinhauses (trans. from Lat. M. Opitz), E, 4vv; 34 Der Esel und die Dohle (Lichtwer), C, 8vv; 35 Schalksnarren (Logau), B♭, 6vv; 36 Zweierlei Feinde (trans. from Arabic A. Tscherning), F/G, 3vv; 37 Der Bäcker und die Maus (Lichtwer), d, 5vv; 38 Die Flinte und der Hase (Lichtwer), G, 4vv; 39 Der Nachbar (Lichtwer), C, 3vv; 4vv; 40 Liebe zur Kunst (Logau), g, 4vv; 41 Frag und Antwort zweier Fuhrleute (Die Welt), g, 5vv; 42 Der Fuchs und der Adler (Lichtwer), ?C, 3vv; 43 Wunsch (Hagedorn), g, 4vv; 44 Gott im Herzen, F (cancelled, incl. in Missa Sancti Bernardi, A 9), 3vv; 45 Turk was a faithful dog (V. Rauzzini), B♭, 4vv; 46 Thy voice o harmony, C, 3/4vv (arr. of no.a: 1); 47 Canon without text, G, 7vv	c1791–9 except nos. 45–6:	A (u)/HV/ HC, Sk — HW xxxi, 21–65; critical commentary, 16 (no.47) and passim (sketches)

Note: canon Der Spiess, listed in Landon, v (A1977), 317, misquoted; *recte* Der Hirsch, no.32

2: INSTRUMENTAL

J: SYMPHONIES

I, II, III, IV = number of movement

HI	Key	Forces	Date	Authentication	Edition	Remarks	
1	D	2 ob, 2 hn, str	~25 Nov 1759 [?1757]	HV, Gr	HW i/1, 1; P i, 37		
2	C	2 ob, 2 hn, str	~1764 [~?1761]	EK	HW i/1, 41; P i, 51	MS copy, A-ST, with spurious Minuet	10

H	Key	Forces	Date	Authentication	Edition	Remarks
3	G	2 ob, 2 hn, str	-1762	EK	P i, 71	
4	D	2 ob, 2 hn, str	-1762 [-?1760]	EK	HW i/1, 59; P i, 89	
5	A	2 ob, 2 hn, str	-1762 [-?1760]	HV, F	HW i/1, 206; P i 107	ed. M i/1 with order of I and II reversed
6	D	fl, 2 ob, bn, 2 hn, str	?1761	HV	HW i/3, 1; P i, 125	'Le matin': title probably authentic [14]
7	C	fl, 2 fl/ob, bn, 2 hn, str	1761	A	HW i/3, 32; P i, 157	'Le midi': title authentic; facs. (Budapest, 1972) [14]
8	G	fl, 2 ob, bn, 2 hn, str	?1761	HV	HW i/3, 73; P i, 197	'Le soir': title probably authentic; IV: 'La tempesta'; I quotes air from Gluck: Le diable à quatre [14]
9	C	2 fl/ob, bn, 2 hn, str	1762	A, EK	HW i/3, 112; P i, 231	? I and II orig. ov. to unidentified vocal work
10	D	2 ob, 2 hn, str	-1766 [-?1761]	HV, F	HW i/1, 90; P i, 243	
11	E♭	2 ob, 2 hn, str	-1769 [-?1760?]	HV, F (? = RC)	HW i/1, 187; P i, 259	
12	E	2 ob, 2 hn, str	1763	A	HW i/3, 146; P i, 279	
13	D	fl, 2 ob, 4 hn, (timp), str	1763	A	HW i/3, 161; P ii, 3	
14	A	2 ob, 2 hn, str	-1764 [-?1762]	HV, JE	P ii, 29	II also used in N 14; for bracketed dates of nos.14, 16–20, 26, 34, 38, 41, 52, 59, 108, see Gerlach (M1996), 19ff
15	D	2 ob, 2 hn, str	-1764 [-?1761]	EK	P ii, 43	
16	B♭	2 ob, 2 hn, str	-1766 [-?1765]	HV	P ii, 65	
17	F	2 ob, 2 hn, str	-1765 [-?1762]	EK	HW i/1, 130; P ii, 79	
18	G	2 ob, 2 hn, str	-1766 [-?1762]	EK	HW i/1, 27; F ii, 97	ed. M i/2 with order of I and II reversed
19	D	2 ob, 2 hn, str	-1766 [-?1762]	EK	HW i/1, 145; P ii, 113	
20	C	2 ob, 2 hn, 2 tpt, timp, str	-1766 [-?1762]	EK	HW i/1, 104; P ii, 127	
21	A	2 ob, 2 hn, str	1764	A	HW i/4, 1; P ii, 155	
22	E♭	2 eng hn, 2 hn, str	1764	A	HW i/4, 15; P ii, 173	'The Philosopher'; another version, Hl:22bis, incl. doubtful Andante grazioso (P ii, 189) [14, 50]
23	G	2 ob, 2 hn, str	1764	A	HW i/4, 31; P ii, 197	
24	D	fl/2 ob, 2 hn, str	1764	A	HW i/4, 48; P ii, 217	
25	C	2 ob, 2 hn, str	1766 [-?1760]	F	HW i/1, 172; P ii, 237	
26	d	2 ob, 2 hn, str	1770 [-?1768]	EK	P ii, 253	'Lamentatione': title ?authentic; title Weihnachtssymphonie (M i/2) of no apparent relevance [18]
27	G	2 ob (2 hn), str	-1766 [-?1761]	EK	HW i/1, 75; P ii, 271	
28	A	2 ob, 2 hn, str	1765	A	HW i/4, 65; P iii, 3	
29	E	2 ob, 2 hn, str	1765	A	HW i/4, 80; P iii, 21	cf R 20
30	C	fl, 2 ob, 2 hn, str	[-?13 Sept] 1765	A	HW i/4, 96; P iii, 41	'Alleluja'; Gregorian Easter Alleluia quoted in I; cf Q 64 [14]
31	D	fl, 2 ob, 4 hn, str	[-?13 Sept] 1765	A	HW i/4, 109; P iii, 57	'Hornsignal'; title 'Auf dem Anstand' (M i²3) of no apparent relevance [14]
32	C	2 ob, 2 hn, 2 tpt, timp, str	-1766 [-?1760]	EK	HW i/1, 223; P iii, 95	

H	Key	Forces	Date	Authentication	Edition	Remarks	
33	C	2 ob, 2 hn, 2 tpt, timp, str	−1767 [−?1760]	EK	P iii, 117	MS copy, *CZ-Bm*, with doubtful Andante	
34	d/D	2 ob, 2 hn, str	−1767 [−?1765]	EK (with incipit of II)	P iii, 143		18
35	B♭	2 ob, 2 hn, str	1 Dec 1767	A	HW i/6, 1; P iii, 165		
36	E♭	2 ob, 2 hn, str	−1769 [?±1761–5]	EK	P iii, 187		10
37	C	? 2 ob, 2 hn (/2 tpt, timp), str	−?1758	EK	HW i/1, 14; P iii, 211		18
38	C	2 ob, 2 hn, (2 tpt, timp), str	−1769 [−?1768]	EK	P iii, 227		
39	g	2 ob, 4 hn, str	−1770 [?1765]	EK	P iii, 253		
40	F	2 ob, 2 hn, str	1763	A	HW i/3, 124; P iii, 277		
41	C	fl, 2 ob, 2 hn, (2 tpt, timp), str	−1770 [−?1768]	EK	P iv, 3	date on MS in *D-Tl*	18
42	D	2 ob, 2 bn, 2 hn, str	1771	A	HW i/6, 43; P iv, 41	'Mercury'	18
43	E♭	2 ob, 2 hn, str	−1772	EK	P iv, 73	'Mourning'; 'Trauersinfonie'	18
44	e	2 ob, 2 hn, str	−1772	EK	P iv, 107		18
45	f♯	2 ob, bn, 2 hn, str	1772	A	HW i/6, 69; P iv, 139	'Farewell'; facs. (Budapest, 1959)	16, 18, 49, 50, 52
46	B	2 ob, 2 hn, str	1772	A	HW i/6, 104; P iv, 175		18, 52
47	G	2 ob, bn, 2 hn, str	1772	A	HW i/6, 125; P iv, 199	cf W 24	18
48	C	2 ob, 2 hn (/2 tpt, timp), str	−?1769	EK	P iv, 233	'Maria Theresa'; facs. of Elssler MS in *CS-Mms*	18
49	f	2 ob, 2 hn, str	1768	A	HW i/6, 24; P iv, 271	'La passione'; 'Il quakuo di bel'humore'	18
50	C	2 ob, 2 hn, 2 tpt, timp, str	1773	A	HW i/7, 1; P v, 3	I and II supposedly composed as ov. to Vorspiel: Der Götterrat (E 12); autograph in *PL-Kj*	19
51	B♭	2 ob, 2 hn, str	−1774	EK	P v, 31	1st of the 2 trios missing in some sources	19
52	c	2 ob, (bn), 2 hn, str	−1774 [−?1772]	EK	P v, 57		18
53	D	3 versions fl, 2 ob, bn, 2 hn, (timp), str	?1778/9	EK (slow introduction)	P v, 97	'Imperial', 'Festino'; 3rd finale (P v, 150; cf H: C, C''), spurious; other combinations dubious (H: D, E', E'')	22
B''	(i)	finale: 2 ob, 2 hn, str		HV	finale: P v, 135	no slow introduction; finale uses concert version of ov, K 3; cf no.62	
B'	(ii)	finale: as (i)			finale: as (i)	as (i), with introduction	
A	(iii)	finale: as other movts		JE	finale: P v, 124	as (ii), with new finale	
54	G	2 fl, 2 ob, 2 bn, 2 hn, 2 tpt, timp, str	1774	A	HW i/7, 28; P v, 163	slow introduction apparently an afterthought; fl, tpt, timp pts added later	19
55	E♭	2 ob, bn, 2 hn, str	1774	A	HW i/7, 63; P v, 201	'The Schoolmaster'	19
56	C	2 ob, bn, 2 hn, 2 tpt, timp, str	1774	A	HW i/7, 86; P v, 229		19
57	D	2 ob, 2 hn, str	1774	A	HW i/7, 126; P v, 271		19
58	F	2 ob, 2 hn, str	−1774 [−?1767/8]	EK	P vi, 3	see Q 52	18
59	A	2 ob, 2 hn, str	−1769 [−?1768]	EK	P vi, 21	'Fire'; cf appx K 1	18
60	C	2 ob, 2 hn, (2 tpt), timp, str	−1774	EK	P vi, 43	'Il distratto'; 'Der Zerstreute'; title authentic; cf K 2	18, 19, 58
61	D	fl, 2 ob, 2 bn, 2 hn, timp, str	1776	A	HW i/8, 175; P vi, 75		22

H	Key	Forces	Date	Authentication	Edition	Remarks	
62	D	fl, 2 ob, (2) bn, 2 hn, str	-1781 [?1780]	EK	P vi, 127	I is rev. version of Finale B from no.53	22
63	C	fl, 2 ob, bn, 2 hn, str	-1781 [?1779]	EK	P vi, 198	'La Roxolane', 'Roxolana': title authentic, refers to II; I is altered version of ov. to Il mondo della luna (E 17); earlier version uses finale of frag. K 1; for Versione prima', see appx K 4	22
64	A	2 ob, 2 hn, str	-1778 [-?c1773]	EK	P vi, 235	'Tempore mutantur': title probably authentic	19
65	A	2 ob, 2 hn, str	-1778 [?c1769–72]	EK	P vi, 259		18
66	B♭	2 ob, 2 bn, 2 hn, str	-1779 [?c1775/6]	EK	HW i/8, 135; P vii, 3		19
67	F	2 ob, 2 bn, 2 hn, str	-1779 [?c1775/6]	EK	HW i/8, 47; P vii, 55		19
68	B♭	2 ob, 2 bn, 2 hn, str	-1779 [?c1774/5]	EK	HW i/8, 1; P vii, 109	order of II and III sometimes reversed as in H; abridged version (HW i/8, 228), doubtful	19
69	C	2 ob, 2 bn, 2 hn, 2 tpt, timp, str	-1779 [?c1775/6]	EK	HW i/8, 93; P vii, 163	'Laudon', 'Loudon': title approved by Haydn; see note to appx X.2	19
70	D	fl, 2 ob, bn, 2 hn, 2 tpt, timp, str	-18 Dec 1779 [?1778/9]	EK, A (timp pr, u)	P vii, 217	timp (? and tpts) added later by Haydn	22, 50
71	B♭	fl, 2 ob, bn, 2 hn, str	-1780 [?1778/9]	EK	P vii, 249		22
72	D	fl, 2 ob, bn, 4 hn, (timp), str	-1781 [?c1763–5]	EK	P vii, 305		72
73	D	fl, 2 ob, 2 bn, 2 hn, (2 tpt, timp), str	-1782 [?1781]	EK, A (u frag.)	P vii, 331	'La chasse': title authentic, refers to IV, orig. composed as ov. to La fedeltà premiata (E 21); II uses song, Gegenliebe (G 16); autograph frag. in ?L-Kj	22
74	E♭	fl, 2 ob, bn, 2 hn, str	-22 Aug [?1781]	EK	P viii, 3		22
75	D	fl, 2 ob, bn, 2 hn, (2 tpt, timp), str	-1781 [?1779]	EK	P viii, 53		22
76	E♭	fl, 2 ob, 2 bn, 2 hn, str	?1782	EK	P viii, 101	nos.76–8 apparently for Haydn's planned visit to England, 1783; see Haydn's letter, 15 July 1783	23, 24, 30
77	B♭	fl, 2 ob, 2 bn, 2 hn, str	?1782	EK	P viii, 153		23, 24, 30
78	c	fl, 2 ob, 2 bn, 2 hn, str	?1782	EK	P viii, 207		23, 24, 30
79	F	fl, 2 ob, 2 bn, 2 hn, str	-?20 Nov 1784	HV, RC	P viii, 255		23
80	d	fl, 2 ob, 2 bn, 2 hn, str	-8 Nov 1784	HV, SC	P viii, 311		23
81	G	fl, 2 ob, 2 bn, 2 hn, str	-8 Nov 1784	EK	P viii, 363		23
nos.82–7: Paris syms.							
82	C	fl, 2 ob, 2 bn, 2 hn/rpt, timp, str	1786	A, A (u frag.)	HW i/13, 107; P ix, 3	'L'ours', 'The Bear'; orig. version of Trio, L i/9, 303, HW i/13, 179	25, 27, 73
83	g	fl, 2 ob, 2 bn, 2 hn, str	1785	A	HW i/12, 91; P ix, 61	'La poule', 'The Hen'	26
84	E♭	fl, 2 ob, 2 bn, 2 hn, str	1786	A, Sk	HW i/13, 1; F ix, 107	sketch for II, HW i/13, 163	26
85	B♭	fl, 2 ob, 2 bn, 2 hn, str	?1785	A (u frag.), HV, SC	HW i/12, 49; P ix, 161	'La reine', 'The Queen [of France]'	26
86	D	fl, 2 ob, 2 bn, 2 hn, 2 tpt, timp, str	1786	A, Sk	HW i/13, 52; P ix, 207	sketches for I and III, HW i/13, 164–5, 168–9	26

H	Key	Forces	Date	Authentication	Edition	Remarks	
87	A	fl, 2 ob, 2 bn, 2 hn, str	1785	A	HW i/12, 1; P ix, 261		
nos.88–9 composed for J. Tost							
88	G	fl, 2 ob, 2 bn, 2 hn, 2 tpt, timp, str	?1787	EK	P x, 3		26
89	F	fl, 2 ob, 2 bn, 2 hn, str	1787	A	P x, 59	II and IV use lira conc., T 4	26
nos.90–92 composed for Comte d'Ogny and Prince Oettingen–Wallerstein							
90	C	fl, 2 ob, 2 bn, 2 hn, (2 tpt, timp), str	1788	A	P x, 109		26, 32
91	Eb	fl, 2 ob, 2 bn, 2 hn, str	1788	A	P x, 167		26, 32
92	G	fl, 2 ob, 2 bn, 2 hn, (2 tpt, timp), str	1789	A	P x, 223	'Oxford'; unpubd version of II and IV for fl, 2 ob, 2 hn, str in later MS authorized copy	26, 31, 32, 52
nos.93–104: London syms.							
93	D	2 fl, 2 ob, 2 bn, 2 hn, 2 tpt, timp, str	1791	A (lost), EK	P xi, 3	perf. London, 17 Feb 1792	32
94	G	2 ob, 2 bn, 2 hn, 2 tpt, timp, str	1791	A	HW i/16, 64; P xi, 49	'The Surprise'; perf. London, 23 March 1792; 1st version of II without 'surprise', HW i/16, 203, P xi, 116; cf C 6	32, 49, 57
95	c	fl, 2 ob, 2 bn, 2 hn, 2 tpt, timp, str	1791	A	P xi, 121	perf. London, 1791	32, 35
96	D	2 fl, 2 ob, 2 bn, 2 hn, 2 tpt, timp, str	1791	A (incl. Sk)	P xi, 171	'The Miracle'; perf. London 1791; sketch for II, P xi, 219; see note to appx X.2	32, 35
97	C	2 fl, 2 ob, 2 bn, 2 hn, 2 tpt, timp, str	1792	A	HW i/16, 122; P xi, 223	perf. London, 3/4 May 1792	32
98	Bb	fl, 2 ob, 2 bn, 2 hn, 2 tpt, timp, hpd obbl, str	1792	A	HW i/16, 1; P xi, 301	perf. London, 2 March 1792	32
99	Eb	2 fl, 2 ob, 2 cl, 2 bn, 2 hn, 2 tpt, timp, str	1793	A, Sk	HW i/17, 1; P xii, 3	perf. London, 10 Feb 1794; autograph in PL-Kj; sketches for Finale, critical commentary to HW i/17, 49a, P xii, 402; cf appx Y.1, 5	35, 36, 51, 53
100	G	(2) fl, 2 ob, 2 cl, 2 bn, 2 hn, 2 tpt, timp, perc, str	1793/4	A	HW i/17, 145; P xii, 59	'Military'; perf. London, 31 March 1794; II uses Romance of lira conc., T 5; cf K 12	35, 36
101	D	2 fl, 2 ob, 2 cl, 2 bn, 2 hn, 2 tpt, timp, str	1793/4	A, Sk	HW i/17, 59; P xii, 139	'The Clock'; perf. London, 3 March 1794; autograph in PL-Kj; sketches for Minuet and Trio, critical commentary to HW i/17, 57a, P xii, 406; cf appx Y.1, 3	35, 36, 49
102	Bb	2 fl, 2 ob, 2 bn, 2 hn, 2 tpt, timp, str	1794	A	HW i/18, 1; P xii, 205	perf. London, 2 Feb 1795; II uses Adagio of pf trio, V 24	36
103	Eb	2 fl, 2 ob, 2 cl, 2 bn, 2 hn, 2 tpt, timp, str	1795	A	HW i/18, 59; P xii, 265	'Drumroll'; perf. London, 2 March 1795; 1st version of closing section, HW i/18, 224, P xii, 326	36, 51, 52

H	Key	Forces	Date	Authentication	Edition	Remarks	
104	D	2 fl, 2 ob, 2 cl, 2 bn, 2 hn, 2 tpt, timp, str	1795	A, Sk	HW i/18, 129; P xii, 333; facs. (Leipzig, 1983)	'London', 'Salomon'; perf. London, 4 May 1795	36, 52
105	Bb	Concertante, soli: vn, vc, ob, bn; fl, ob, bn, 2 hn, 2 tpt, timp, str	1792	A (incl. Sk)	HW ii; P x, 287	perf. London, 9 March 1792; sketch for I, HW ii, 72, P x, 371	34
106	D	2 ob, 2 hn, str	?1769	EK	HW xxv/4, 289	only I extant, as III of K 5; supposedly composed as ov. to Le pescatrici (E 10); cf K 5	
107	Bb	2 ob, 2 hn, str	-1762 [-?1761]	F	HW i/1, 158; P i, 3	syn. 'A'; cf appx O.3, 1; also attrib. Wagenseil syn. 'B'	
108	Bb	2 ob, bn, 2 hn, str	-1765	HV	P i, 19	syn. 'B'	

Note: single movrs, see K 1, 10; c150 spurious syms. listed in H

K: MISCELLANEOUS ORCHESTRAL

No.	H	Title, key	Forces	Date	Authentication	Edition	Remarks	
1	i, 87	Menuet, Trio, Finale, C	2 ob, 2 hn, 2 tpt, timp str	?1773	A (u frag.)	P vi, 184	finale used for earlier version of sym. J 63; cf no.8	
2	XXX:3	Incidental music: Der Zerstreute (comedy, 5, ? J.B. Bergopzoomer, after J.F. Regnard: Le distrait)	see J 50	-30 June 1774	Pressburger Zeitung, 23 Nov 1774	see J 60	ov., entr'actes and final music = sym. J 60	
3	Ia:7	Sinfonia, D	2 ob, 2 bn, 2 hn, str	1777	A, A (frag.)	P v, 135	1 movt only; ov. to unidentified work (= ? appx E 1); used as Finale B of sym. J 53; cf appx K 5	22
4–9		6 sinfonie [overtures]:						
4	Ia:13	g	fl, 2 ob, bn, 2 hn, str	-29 Sept 1782	OE	(London, 1959)	ov. to L'isola disabitata (E 20)	
5	Ia:6	D	2 ob, 2 hn, str			HW xxv/6, 1	ov. to L'incontro improvviso (E 14); tpts, timp, perc omitted; ?earlier version uses I of J 106 for III	
6	Ia:10	G	fl, 2 ob, 2 hn, str			HW xxv/3, 1	ov. to Lo speziale (E 9)	
7	Ia:15	Bb	2 ob, bn, 2 hn, str			HW xxv/8, 1	ov. to La vera costanza (E 19); with added III compiled from Introduzione of same opera and balletto from Il mondo della luna (E 17)	
8	Ia:1	C	2 ob, 2 hn, timp, str		A (u frag.)	HW xxv/5, 1	ov. to L'infedeltà delusa (E 11); with altered II and added III almost identical with III of ov. to Die Feuersbrunst (E 16b); ?earlier version with altered II, uses no.1 as III and IV	
9	Ia:2	c/C	2 ob, 2 bn, 2 hn, 2 tpt, timp, str			HW xxviii/-, 1	ov. to Il ritorno di Tobia (C 3); with altered final bars	

No.	H	Title, key	Forces	Date	Authentication	Edition	Remarks
10	Ia:4	Finale, D	fl, 2 ob, 2 bn, 2 hn, str	?1777–86 [?1782–4]	A, EK	D 51 (1959)	from unidentified work (? sym. J 73); sometimes connected with sym. J 53
11	XX/1 A	Musica instrumentale sopra le 7 ultime del nostro Redentore in croce, ossiano 7 sonate con un'introduzione ed al fine un terremoto (The Seven Last Words)	2 fl, 2 ob, 2 bn, 4 hn, 2 tpt, timp, str	–11 Feb 1787 [?1786]	OE, Sk, RC	HW iv	composed for Cádiz; ?1st Viennese perf. 26 March 1787; some sketches in critical commentary to HW iv, 41; cf C 4, appx O.1, 1; see note to appx X.2 [26, 45]
12	i, 206	Piece for military band, C	fl, 2 ob, 2 cl, 2 bn, 2 hn, tpt, serpent, perc	?1794/5		A (u)	HW i/17, 227 arr. of II of sym. J 100
13	—	Overtura Conventgarden	?	1791–5	Gr, Dies	—	lost or = HIa:3, ov. to Orfeo (E 24); ?perf. as ov. to J.P. Salomon's opera Windsor Castle; cf appx K 6
14	i, 590	?, E	?	?c1763–9	A (u)	—	b only of sequence of 9 pieces; see critical commentary to HW xiii, 12

Note: ovs. listed as HIa:3, 5, 8, 14, 16, 17 Add. are taken from Haydn's operas without alteration; Ia:11 is not a separate piece

Appendix K: Doubtful and spurious works or arrangements

No.	H	Title, librettist, key	Forces	Date	Edition	Remarks
1	XXX:2	Incidental music: Die Feuersbrunst (? G.F.W. Grossmann)	?	?1773/4	—	not verified; identical neither with E 16a/b nor with sym. J 59
2	XXX:B	Incidental music: Hamlet (W. Shakespeare)	?	?c1774–6	—	not verified
3	XXX:C	Incidental music: Götz von Berlichingen (J.W. von Goethe)	?	–?1776	—	not verified; also attrib. M. Haydn
4	XXX:D	Incidental music: Soliman II, oder Die drei Sultaninnen (? F.X. Huber, after C.-S. Favart)	?	?c1777	—	not verified; hypothetical reconstruction (P vi, 165; '1st version' of sym. J 63) combines altered ov. to Il mondo della luna (E 17), II of J 63 and frag. K 1
5	Ia:7bis	Overture, D	?	–1783	—	2nd version of K 3; extant in various arrs. only; inclusion in sym. J 53 doubtful
6	ii, 435	Overture to Salomon's opera Windsor Castle, D	?	1795	—	in MS (*J-Tn*) and in vs of opera (London, n.d.), as by Salomon; cf K 13
7	XXX:A; Ia:9	Incidental music: King Lear (Shakespeare)	2 fl, 2 ob, bn, 2 hn, 2 tpt, timp, str	–1806	—	ov. and entr'actes, without author's name; ov. attrib. W.G. Stegmann (? C.D. Stegmann); ? by J. von Blumenthal (i)
8	Ia:12	Overture, g	2 ob, bn, 2 hn, str	–1799	—	combination and arr. of pieces from Il mondo della luna (E 17)

No.	E	Title, librettist, key	Forces	Date	Remarks
		Fantaisie, d			see appx X.3, 16

E: DANCES, MARCHES FOR ORCHESTRA/MILITARY BAND

No.	H	Title, key	Forces	Date	Authentication	Edition	Remarks
1	IX:1	[12] Minuetti (with 3 Trios)	2 ob, 2 hn, 2 vn, b	?-760	A	HW v, 3; D 855 (1983)	'Seitenstetten' minuets
2	IX:3	[12] Menuetti (with 4 Trios)	(?2) fl pic, (?2) fl, 2 ob, (?2) bn, 2 hn 2 vn, b	?-767	—	—	lost; pf arr. extant, cf appx X.2, 1
3	—	4 [?cycles of] Menuetti	?	~1765	EK	—	? partly = nos.1 and 2; otherwise lost
4	iii, 315	Marche regimento de Marshall, G	2 ob, 2 hn, 2 bn	~1772	—	HW v, 218, D 34, 2 (196C)	—
5	IX:23	?24 Dances (?12 Minuets and 12 Trios)	2 fl, 2 hn, 2 vn, b and ?	~?4773	A (frag.)	HW v, 12	only nos.23–4 extant; ? perf. Bratislava, 16 Nov 1772
6a	IX:5	[6] Menuetti (with 2 Trios)	fl, 2 ob, bn, 2 hn, 2 vn, b	1776	A	HW v, 14	? = 1st pt of longer cycle; cf appx L5
6b	—	Menuetti	?	~Aug 1776	Schellinger's account of transcripts	—	? lost; ? = no.6a
7	IX:6a Add.	12 Menuets	?	-11 Feb [-?9 Jan] 1777	see Thomas (M1973)	—	lost or ? identical with no.6; for the Redoutensäle, Vienna
8	IX:6b Add.	18 Menuets	?	-8 Feb [-?9 Jan] 1780	see Thomas (M1973)	—	lost or unidentified; for the Redoutensäle, Vienna
9	IX:7	Raccolta de' [14] menuetti ballabili (with 6 Trios)	fl, 2 ob, 2 bn, 2 hn, timp, 2 vn, b	-31 Jan 1784	—	HW v, 20; D 301 (1970)	—
10	IX:8	XII Menuets (with 5 Trios)	?	-12 Jan 1785	—	—	lost, pf arr. extant, cf appx X.2, 3
11	IX:9	6 Allemandes (6 deutsche Tänze)	2 cb, bn, 2 hn, 2 tpt, timp, 2 vn, b	-15 Nov 1786 [-?19 Jan 1785]	—	HW v, 37; D 52 (1960)	for dates see H iii, 313, HW v, preface
12	IX:9d, e Add.	Unos 24 minués y orras tantas [= 24] contradanzas	?	-22 April 1789	see Solar-Quintes (E1947)	—	sent to Duchess of Osuna (Madrid); lost or unidentified; minués = ?no.14
13	IX:9c	12 ganz neue Tanz Menuetts mit 12 Trios begleitet	?	-1? Jan 1790	Haydn's letters	—	promised to Prince Oettingen-Wallerstein, 21 Oct 1789; lost or unidentified
14	IX:16	24 Menuetti (with 24 Trios)	fl pic, 2 fl, 2 ob, 2 cl, 2 bn, 2 hn, 2 tpt, timp, perc, 2 vn, vc, b	?c1790-1800	—	HW v, 92; D 299 (1974)	—
15	VIII:6	Marcia, E♭	2 c., 2 hn, 2 bn	-1793 [?c1780–90]	A (u)	HW v, 207; xxv/12, 315; D 34, 4 (1960)	cf appx Y3, 3

No.	H	Title, key	Date	Forces	Authentication	Edition	Remarks	
16	VIII:7	March, Eb	?c1792	2 cl, 2 bn, 2 hn, tpt, serpent	A (u frag.)	HW v, 208; D 34, 5 (1960)	only 1st 8 bars extant	
17a	VIII:3	March, Eb	1792	2 cl, 2 bn, 2 hn, tpt, serpent	A (as pt of no.17b); Sk	HW v, 209; D 34, 6 (1960)	? = March for the Prince of Wales mentioned by Gr and Dies	
17b	VIII:3bis	March, Eb	1792/5	2 fl, 2 cl, 2 bn, 2 hn, 2 tpt, str	A	HW v, 220; D 98 (1961)	2nd version of no.17a; for Royal Society of Musicians	
18	IX:11	[12] Menuetti di ballo (Redout Menuetti; Katharinentänze) (with 11 Trios)	[–25 Nov] 1792	fl pic, 2 fl, 2 ob, 2 cl, 2 bn, 2 hn, timp, 2 vn, b	A (u pr), Sk	HW v, 42; (Lippstadt, 1959)	for Redoute of Viennese Pensionsgesellschaft bildender Künstler; sketches in HW v, 180; for pf arr, see appx X.2, 4	35, 66
19	IX:12	12 deutsche Tänze (Tedeschi di ballo) (with Trio and Coda)	[–25 Nov] 1792	2 fl, 2 ob, 2 cl, 2 bn, 2 hn, 2 tpt, timp, 2 vn, b	Sk	HW v, 77; (Lippstadt, n.d.)	for Redoute as above; sketches in HW v, 183; for pf arr, see appx X.2, 5	35, 66
20	—	24 Minuets and German Dances	1791–5	?	Gr, Dies	—	lost or ? = nos.18 and 19	
21	iii, 323	4 and 2 Countrydances	1791–5	?	Gr, Dies	—	lost; for ? pf arr. of one or two, see X 7, appx X.3, 6	
22	VIII:1–2	2 [Derbyshire] Marches, Eb, C	1795	2 cl, 2 bn, 2 hn, tpt, serpent, ?perc	A (nos.1, 2), A (u) (no.2)	HW v, 212; D 34, 8–9 (1960)	for pf arr, see appx X.2, 7	
23	VIII:4	Hungarischer National Marsch, Eb	[–27 Nov] 1802	2 ob, 2 cl, 2 bn, 2 hn, tpt	A	HW v, 216; D 34, 10 (1960)	42	
24	i, 541	March, Eb	after 1791	str	?A	—	lost	

Note: sketches to unknown minuets/Ger. dances in HW v, 180 (transcr.), 250 (facs.); 3 unidentified minuets (with 3 trios) composed for Haydn by J. Eybler in 1789

Appendix L: Selected doubtful and spurious works or arrangements

No.	H	Title	Forces	Date	Edition	Remarks
1	i, 547	VI Menuets (with 6 Trios) and VI Allemandes	(fl, ob), 2 hn, 2 vn, b	–1787	(Berlin and Amsterdam, 1787)	by Haydn and Vanhal; minuets by Vanhal; allemandes identical with L 11
2	i, 547	12 Contratänze	fl, 2 ob, 2 bn, 2 hn, 2 vn, b	–1799	—	lost; ?arr. from various works by Haydn; see Thomas (M1973)
3	IX:2	VI Menuetti	2 hn, 2 vn, b	–?1766	—	lost
4	IX:4	[12] Minuetti da ballo (with 12 Trios)	2 fl, 2 hn, 2 vn, b	–1766	(Amsterdam, 1766)	
5	IX:6, nos.1–8	XII Menuetti (with 4 Trios)	fl, 2 ob, 2 bn, 2 hn/rpt, timp, 2 vn, b	?	—	hIX:6, nos.9–12 = hIX:5, nos.1–4; see L 6
6	IX:9b	12 Deutsche (dell'opera L'arbore di Diana)	fl, 2 ob, cl, bn, 2 hn, 2 vn, b	1787–99	—	lost; pf arr. ? extant as appx X.3, 4
7	IX:14	13 Menuetti (with 4 Trios)	fl pic, 2 fl, 2 ob, 2 bn, 2 hn, 2 vn, b		—	

No.	H	Title	Forces	Date	Edition	Remarks
8	IX:15	[6] Menuetti (with 6 Trios)	2 fl, 2 ob, 2 cl, 2 bn, 2 hn, 2 tpt, timp, 2 vn, b	?	—	
9	IX:17	[17] Deutsche Tänze	fl pic, fl, 2 ob, 2 bn, 2 hn, 2 vn, b	?	—	lost
10	IX:18	IX Menuette (with Trios) fürs Orchester	?	?	—	1st incipit = L 10, no.7, rest unknown
11	IX:19	[13] Menuetti (with 4 Trios)	2 vn, b	~?1777	HW v, 199	theme of no.1 similar to III of S 3; no.11 uses III of S 6
12	IX:24	Menuetto and Trio	2 vn, b	?	—	unsigned draft MS not Haydn's autograph; for pf arr. see appx X.3, 13; for orch arr. see no.14
13	IX:25	Minuet	str	?	—	movt of sym. by Dittersdorf (Krebs no.35)
14	i, 580	10 Menuette	orch	?	(Kassel, 1950)	orch arr. of no.12 and minuets arr. from syms. and str qts
15	i, 580	12 deutsche Tänze, 2 versions	i orch ii 2 vn, vc	?	i (Kassel, 1950); ii HM. xii (1967?	minuets arr. from syms. and str qts
16	—	12 Menuette	2 vn, ?vc	?	(Wolfenbüttel, 1938?	from str trios, baryton trios and Scherzandi
17	iii, 323	XII [recte XIV] Menuette (with 5 Trios)	2 fl pic/fl,ob, 2 hn, 2 vn, b	?	—	see Landon (B1959), 70
18	IX:22a Add.	[12] Menuetti (with 1 Trio)	2 ob, 2 hn, 2 vn, b	?	—	see Thomas (M1973), 23

Note: for doubtful works extant only in pf arrs. see appx X.3

M: CONCERTOS FOR STRING OR WIND INSTRUMENTS

No.	H	Title, key	Forces of orchestral accompaniment	Date	Authentication	Edition	Remarks
1–4	VIIa	[4] Concerti per il violino:					
1	1	C	str	~1769 [?c1761–5]	EK	HW iii²1, 1	for Luigi [Tomasini] 66
2	2	D	(2 ob, 2 hn), str	?c1761–5	EK	—	lost; incipit in HW iii²1, VI
3	3	A (Melker Konzert)	str	~1771 [?c1765–70]	EK	HW iii²1, 32	title not authentic
4*	4	G	str	~1769	—	HW iii²1, 71	
5–7	VIIb	[2/3] Concerti per il violoncello:					
5	1	C	2 ob, 2 hn, str	?c1761–5	EK	HW iii²2, 1	14, 66
6	2	D	2 ob, 2 hn, str	1783	A	HW iii²2, 57	erroneously attrib. A. Kraft; rev. version by F.A. Gevaert
7	3	C	?	?c1761–5	EK	—	lost or = no.5
8	VIIc:1	Concerto per il violone (contraviolone), D	?	?1763	EK	—	lost 14
9–10	XIII	[2] Concerti per il pariton [baryton]:					
9	1	D	(? 2 vn, b)	?c1765–70	EK	—	lost
10	2	D	(? 2 vn, b)	?c1765–70	EK	—	lost
11	XIII:3	Concerto per 2 pariton, D	?	?c1765–70	EK	—	lost
12	VIIf:1	Concerto per il flauto, D	?	?c1761–5	EK	—	lost

No.	H	Title, key	Forces of orchestral accompaniment	Date	Authentication	Edition	Remarks
13	i, facs., V	Concert für Fagott	?	?	see remark	—	lost; mentioned in Haydn's short work-list, c1803–4
14–15	VIId	[2] Concerti per il corno di caccia:					
14	1	D	?	?c1761–5			lost
15	3*	D (no.1)	2 ob, str	1762	EK	HW iii/3, 1	
16	VIId:2	Concerto a 2 corni, Eb	?	–?1784	HV	—	lost
17	VIIe:1	Concerto per il clarino, Eb	2 fl, 2 ob, 2 bn, 2 hn, 2 tpt, timp, str	1796	A	HW iii/3, 23	38, 66

Note: concs. for 2 lire organizzate, see group T; Concertante, see J 105; conc. for vn, org/hpd, see U 3; ? conc. for vn planned in 1799, see Landon (A1976)

Appendix M: Selected doubtful and spurious works

No.	H	Title, key	Forces of orchestral accompaniment	Date	Edition	Remarks
1	VIIa:D1	Violin concerto, D	2 ob, 2 hn, str	–c1777	(Paris, c1777), as by Stamitz	by C. Stamitz
2	VIIa:G1	Violin concerto, G	str	–1771	—	? by M. Haydn
3	VIIa:A1	Violin concerto, A	?	–c1777	(Paris, c1777), as by Giornovichi	by Giornovichi
4	VIIa:B1	Violin concerto, Bb	str	1760	D 3 (1960), as by M. Haydn	by M. Haydn
5	VIIa:B2	Violin concerto, Bb	str	–1767	(Leipzig, 1915)	by Christian Cannabich
6	VIIb:4*	Cello concerto, D	str	–1772	(Leipzig, 1894)	also attributed (?G.B.) Costanzi
7	VIIb:5*	Cello concerto, C	2 fl, 2 ob, 2 cl, 2 bn, 2 hn, str	?c1899	(Berlin, 1899)	'nach einer Skizze ausgeführt und herausgegeben von David Popper'; sketch never found
8	VIIb:g1	Cello concerto, g	str	–1773	—	lost
9	VIIf:D1	Flute concerto, D	str	–1771	(Munich, 1955)	by L. Hofmann
10	VIIg:C1	Oboe concerto, C	2 ob, 2 hn, 2 tpt, timp, str	?c1800	(Wiesbaden, 1954)	orig. attrib. 'H...r'; Haydn's name added later
11	VIId:4	Horn concerto (no.2), D	str	–1781	(London, 1954)	*D-HR*, orig. without author's name; 'par Michael Heiden' added later
12	—	Concerto for 2 horns, Eb	2 ob, 2 hn, str	?	(Amsterdam, 1966)	

Note: 3 concs., 1/2 cl, attrib. Haydn in D. Klöcker: disc notes, *Orfeo*, C 448971A, not authentic

N DIVERTIMENTOS ETC. FOR 4 + STRING AND/OR WIND INSTRUMENTS

string quartets; works with baryton, lira organizzata excepted

No.	HII	Title, key	Forces	Date	Authentication	Edition	Remarks	
1–4		[4] Divertimentos (Cassations) a 9:						69
1	9	G	2 ob, 2 hn, 2 vn, 2 va, b	–1754	EK	HW viii/1, 15		
2	20	F	+ (bn)	–1753 [–?1757]	EK	HW viii/1, 37; D 56 (1962)		
3	17	C	2 cl instead of 2 ob	–c1765	EK	HW viii/1, 80; D 23 (1960), ed. H. Steppan		
4	G1	G		–1758 [?c1760?]	—	HW viii/–, 63; D 47 (1959)	also attrib. M. Haydn (Perger no.114); arr. a 5 without IV (Copenhagen, 1953)	
5–10		[6] miscellaneous works:						
5	24	[V] Variations on a minuet, Eb	fl, 2 eng hn, bn, 2 hn, vn solo, 2 vn, vc, vle	?1761	A (u)	HW viii/2, 99	? movt of larger work	
6	2	Divertimento (Cassation) a 5, G	2 vn, 2 va, b	–1763 [?1753/4]	EK	HW viii/1, 1; D 894 (1988?)		8, 9, 69
7	10	Divertimento a 6 (Der [verliebte] Schulmeister), D	?	–c1755	EK	—	lost	
8	13	Divertimento (? a 6), D	?	–c1755	EK	—	lost	
9	8	Divertimento (Cassation) [a 7]	2 fl, 2 hn, 2 vn, b	–176?	EK	HW viii/1, 129		69
10	D22 Add.	Cassation, D	4 hn, vn, va, b	?c1763	—	D 56 (1960)	considered not authentic in HW viii/1	
11–12		[2] Divertimentos (Cassations) a 6	2 hn, 2 vn, va, b					
11	21	Eb		–1760 [–?1761]	EK	HW viii/1, 107	for spurious arrs. as str qts, see appx O.3, 2–3	69
12	22	D		–1764 [–?1760]	EK	HW viii/1, 118	most sources incl. added variations of 2nd trio (HW viii/1, 207)	69
13–14		[2] Divertimentos (Cassations) [a 6]:	fl, ob, 2 vn, vc, db					
13	1	G		–1763	EK	HW viii/1, 145; D 846 (1984)	for spurious arr. with lute, see Crawford (N1930)	69
14	11	C (Der Geburtstag)		–1763	EK	HW viii/1, 161; D 57 (1961) D 71–6 (1961)	II: 'Mann und Weib'; cf also J *4	69
15–20		6 Scherzandos (Sinfonias, Divertimentos):	fl/2 ob, 2 hn, 2 vn, b	–1765			other versions probably spurious	66
15	33	F			—	HW viii/2, 63		
16	34	C			—	HW viii/2, 67		
17	35	D			—	HW viii/2, 73		
18	36	G			—	HW viii/2, 79		

No.	hII	Title, key	Forces	Date	Authentication	Edition	Remarks	
19	37	E			—	HW viii/2, 85		
20	38	A			EK	HW viii/2, 91		
21–6		[6] Divertimentos [a 6]: 2 ob, 2 hn, 2 bn						
21	15	F (Parthia)		1760	A	HW viii/2, 18; D 29 (1959)	facs. of autograph in *Haydn Yearbook 1962*, 257	[10], [69]
22	23	F (Parthia)		–1765 [?1760]	A (u frag.)	HW viii/2, 24; D 30 (1959)	added movt of doubtful authenticity, D 30, 8	[69]
23	7	C (Feld-Parthie)		–1765	EK	HW viii/2, 12; D 31 (1959)		[69]
24	3	G (Parthie)		–1766	EK	HW viii/2, 6; D 84 (1960)		[69]
25	D18	D (Cassation)		–1765 [?c1760]	—	HW viii/2, 31; D 33 (1959)	in HW viii/2 as of doubtful authenticity	
26	B♭3 = G9 = C12 Add.	B♭/G/C (Parthia)		–1766 [?c1760]	—	HW viii/2, 38, in B♭; D 85 (1960), in G	in HW viii/2 as of doubtful authenticity	
27–32		[6] Divertimentos [a 4–8]:						
27	16	F (Feld-Parthie)	2 eng hn, 2 hn, 2 vn, 2 bn	1760	A	HW viii/2, 43		[69]
28	12	E♭ (Feld-Parthie) (? a 6)	(?2) eng hn, and ?	–1765	EK	—	lost; probably not in B♭ as ^{H}	
29	20bis	A (Feld-Parthie)	?	–?c1768	EK	—	lost	
30	14	C	2 cl, 2 hn	1761	A	HW viii/2, 3; D 32 (1959)		
31	4	F (?D), a 5	2 cl, 2 hn, bn	–?c1768	EK	—	lost; version for 2 hn, baryton, va, b extant; cf R 26; reconstruction, D, by R. Hellyer (Cardiff, 1983)	
32	5	F (?D), a 5 (? a 4)	2 cl, 2 hn, (?bn)	–?c1768	EK	—		[69]

Note: for arrs. of works composed for baryton or lira organizzata, see groups R and T; 2 Divertimenti a più voci (Gr, Dies) probably = London versions of T 8 and T 13

Appendix N: Selected doubtful and spurious works

No.	hII	Title, key	Forces	Date	Edition	Remarks
1	18	Divertimento (Notturno), D	fl, 2 hn, vn, va, b	?	—	probably by Vanhal, though incipit in HV
2	19	Divertimento (Notturno), G	fl, (2 hn), vn, va, b	?	—	probably by Vanhal, though incipit in HV
3	24a	Minuet with variations, D	2 fl, 2 ob, (bn), 2 hn, 2 vn, va, b	?	—	lost; minuet from P 15
4	24b	Minuet with variations, A	2 fl, 2 ob, 2 hn, 2 vn, va, b	?	—	lost; minuet from P 7
5	39	Divertimento (Echo), E♭	4 vn, 2 b	–1766/7	(Wilhelmshaven, 1957)	
6	40	Sextetto, E♭	ob, bn, hn, vn, va, b	–1781	(London, 1957)	
7–12		6 Divertimentos (Feldparthien):				
7	46	B♭	2 ob, 2 hn, 3 bn, serpent		(New York, 1960)	II: St Antony chorale, basis for Brahms's Haydn variations, op.56; arr. for 5 wind insts (London, 1942)

No.	hII	Title, key	Forces	Date	Edition	Remarks
8	42	Bb	2 ob, 2 cl, 2 hn, 2 bn		—	various versions extant; sometimes without author's name, sometimes with names of M. Haydn, L. Mozart, or E. Angerer; version by L. Mozart with added movts
9	41	Eb	2 ob, 2 cl, 2 hn, 2 bn		(Vienna, 1931)	
10	45	F	2 ob, 2 hn, 3 bn, serpent		—	
11	43	Bb	2 ob, 2 cl, 2 hn, 2 bn		(Mainz, 1970)	probably by L. Hofmann
12	44	F	2 ob, 2 hn, 3 bn, serpent			
13	47	Toy Symphony (Kinder-sinfonie; Berchtolsgadener Musik/Divertimento/ Sinfonie; Symphonie burlesque), C	vn, va (or 2 vn), b, children's insts	~1786	D 300 (1974), in 'Cassatio ex G' by L. Mozart; in Opera incerta (Mainz, 1991), ed S. Gerlach, as by ?M. Haydn	
14	D5	[12] Notturni (Quartetto), D	2 fl, 2 hn	?	(Leipzig and Berlin, 1952)	
15	D6	Divertimento, D	fl, vn, va, b	~1766	(Frankfurt, 1971)	
16	D8	Divertimento (Quintetto), D	fl, 2 vn, va, b	~?1778	(Zürich, 1940)	
17	D9	Quatuor, D	fl, vn, va, b	~1768	(London, 1960)	
18	F2	Cassation, F	cb, bn, 2 hn, vn, va, b	?	(Leipzig, 1970)	
19	F7	Parthia (Harmonie; Octett), F	2 ob, 2 cl, 2 hn, 2 bn	~1802	(Leipzig, 1902)	probably by P. Wranitzky
20	F10	Quartetto, F	3 vn, vc	~1799	(Augsburg, c1799), as by Ferandini	probably by Ferandini; see Marrocco (H1972)
21	F12	Parthia, F	2 ob, 2 hn, bn	?	(London, 1961)	
22	G4	Quatuor, G	fl, vn, va, b	~1768	(London, 1960)	
23	A1	Divertimento, A	2 vn, 2 va, b	~1762	—	considered authentic by Landon (A1980)
24	B4	Divertissement, Bb	ob, vn, b viol, b	?	(London, 1929), (Munich, 1972)	supposed autograph not authentic; also attrib. C.F. Abel and J.C. Bach
25	—	Quatro, C	fl, vn, va. vc	?	(Zürich, 1969)	see HW viii/1, 224, no.JHI-C13
26	D23 Add.	Divertimento, D	2 ob, 2 cl, 2 hn, 2 bn	?	D 86 (1960)	considered authentic by Fruehwald (H1988)

Note: more doubtful and spurious works listed in hII; see also HW viii/1, 221, v ii/2, 120; for arrs. for lute/gui, vn, va, vc, see O 8

O: STRING QUARTETS

No.	hIII	Op.	Title, key	Date (pubd)	Authentication	Edition	Remarks	
1–10			[10] Divertimentos (Cassations, Notturnos):	(1764–6)			for proposed earlier dating see text, §2	
1	1	1/1	Bb	~1762 [?c1757–9]	EK	HW xii/1, 1	'La chasse' for spurious art. for lute, vn, vc, see Crawford (N1980)	9
2	2	1/2	Eb	~1762 [?c1757–9]	EK	HW xii/1, 9		9
3	3	1/3	D	~1762 [?c1757–9]	EK	HW xii/1, 18		9
4	4	1/4	G	~1764 [?c1757–9]	EK	HW xii/1, 27		9
5	II:6	1/0	Eb	~1764 [?c1757–9]	EK	HW xii/1, 39		9

No.	hIII	Op.	Title, key	Date (pubd)	Authentication	Edition	Remarks	
6	6	1/6	C	–1762 [?c1757–9]	HV, F	HW xii/1, 50	arrs. for lute/gui, vn, vc, not authentic	9
7	7	2/1	A	–1763 [?c1760–62]	EK	HW xii/1, 59		9
8	8	2/2	E	–1765 [?c1760–62]	EK	HW xii/1, 69	arrs. (in D) for lute/gui, vn, va, vc, not authentic	9
9	10	2/4	F	–1762 [?c1760–62]	EK	HW xii/1, 80		9, 50
10	12	2/6	Bb	–1762 [?c1760–62]	EK	HW xii/1, 91		9
11–16			6 Divertimentos:	–1771 [?1769/70] (1771/2)	EK			18, 67–8
11	22	9/4	d			HW xii/2, 3		
12	19	9/1	C			HW xii/2, 13		
13	21	9/3	G			HW xii/2, 24		
14	20	9/2	Eb			HW xii/2, 35	cf X 3	
15	23	9/5	Bb			HW xii/2, 45		
16	24	9/6	A			HW xii/2, 57		
17–22			6 Divertimentos:	1771 (1772)	A			18, 67–8
17	26	17/2	F			HW xii/2, 69		
18	25	17/1	E			HW xii/2, 84		
19	28	17/4	c			HW xii/2, 99		
20	30	17/6	D			HW xii/2, 115		
21	27	17/3	Eb			HW xii/2, 129		
22	29	17/5	G			HW xii/2, 140		
23–8			6 Divertimentos:	1772 (1774)	A			18, 67–8
23	35	20/5	f			HW xii/3, 3	'Recitative'	50
24	36	20/6	A			HW xii/3, 21	'Sun Quartets'	50
25	32	20/2	C			HW xii/3, 36		50, 52
26	33	20/3	g		Sk	HW xii/3, 54, 191 (incl. sketch for III)		
27	34	20/4	D			HW xii/3, 70		
28	31	20/1	Eb			HW xii/3, 89		
29–34			6 Quatuors (Quartetti):	1781 (1782)	HV, OE, SC (title, frag. of no.29)		'Russian' Quartets, 'Jungfernquartette', 'Gli scherzi'; date 1778–81 is incorrect	23, 27, 68–9
29	41	33/5	G		C	HW xii/3, 105	'How do you do?'; for pf arr. of IV, see appx X.2, 2	
30	38	33/2	Eb		C	HW xii/3, 120	'The Joke'	49
31	37	33/1	b		C	HW xii/3, 133		49
32	39	33/3	C			HW xii/3, 147	'The Bird'	
33	42	33/6	D		C	HW xii/3, 163		
34	40	33/4	Bb			HW xii/3, 175		
35	43	42	Quartetto, d	1785 (1786)	A	(Vienna, 1988)		24
36–41			6 Quartetti:	1787 [–16 Sept] (1787)	HV, OE, SC	(Vienna, 1985)	'Prussian' Quartets; autographs of nos.38–41 in private collection	25, 69

No.	HIII	Op.	Title, key	Date (publ.)	Authentication	Edition	Remarks	
36	44	50/1	B♭					25, 49, 52, 69
37	45	50/2	C					69
38	46	50/3	E♭					
39	47	50/4	f♯		A			
40	48	50/5	F		A		II: 'Ein Traum'	
41	49	50/6	D		A		'The Frog'	
42–7			6 Quatuors:	–22 Sept 1788 (1789, 1790)	HV, ?Haydn's letters	(Vienna, 1986–7)	'Tost' Quartets, 1st ser.	
42	57	54/2	C					
43	58	54/1	G		A (u frag.)		cf appx Y.1, 1	
44	59	54/3	E		A (u frag.)			
45	60	55/1	A					
46	61	55/2	f				'The Razor'	25
47	62	55/3	B♭					
48–53			6 Quartetti:	1790 (1791)			'Tost Quartets, 2nd ser.	25, 31, 69
48	65	64/1	C		A	HW xii/5, 3		
49	68	64/2	b		A	HW xii/5, 18		
50	67	64/3	B♭		A	HW xii/5, 33		
51	66	64/4	G		HV	HW xii/5, 53		
52	64	64/6	E♭		A	HW xii/5, 68		
53	63	64/5	D		A, Sk	HW xii/5, 83, 218 (incl. sketch for I)	'The Lark'; cf appx Y.1, 4	
54–9			6 Quartetti:	1793 (1795, 1796)			'Apponyi' Quartets	35, 69
54	69	71/1	B♭		A	HW xii/5, 101		
55	70	71/2	D		Sk	HW xii/5, 119, 222 (incl. sketch for III)	cf appx Y.1, 2	
56	71	71/3	E♭			HW xii/5, 135		
57	72	74/1	C			HW xii/5, 155		
58	73	74/2	F		Sk	HW xii/5, 177, 220 (incl. sketch for IV)		
59	74	74/3	g	–14 June 1797 (1799)	HV, OE	HW xii/5, 198	'The Rider'; for pf arr. of II, see appx X.2, 6	52
60–65			6 Quartetti:				'Erdödy' Quartets	38, 69
60	75	76/1	G		HE (= E)			
61	76	76/2	d		Sk		'Fifths'	
62	77	76/3	C	–28 Sept 1797	HE (= E)		'Emperor'; II uses Gott erhalte Franz den Kaiser, G 43; for pf arr. cf appx X.2, 8; facs. of sketch for II (Graz, 1982, 2/1995)	39
63	78	76/4	B♭		HE (= E)		'Sunrise'	
64	79	76/5	D					
65	80	76/6	E♭		HE (= E)			

No.	hIII	Op.	Title, key	Date (pubd)	Authentication	Edition	Remarks	
66–7			2 Quartetti:	1799 (1802)	A	(Vienna, 1982–4)	'Lobkowitz' Quartets; facs. (Budapest, 1972, 2/1980)	69
66	81	77/1	G					38
67	82	77/2	F					38
68	83	103	Unfinished Quartet, d (not B♭)	~1803 (1806)	A, Sk	(Vienna, 1982)	movts II and III only	42

Note: Haydn apparently wrote ?2 small str qts for Spain in 1784, now lost, not identical with hIII:B4, G5, C8, F2, D2 or g 1, which are by Gallus-Mederitsch; unpubd sketch, d, c1795, not identified

Appendix O.1: Arrangements for string quartet

No.	H	Title, key	Date	Authentication	Edition	Remarks	
1	III:50–56	Musica instrumentale sopra le 7 ultime parole del nostro Redentore in croce … ridotte in quartetti, op.51	~14 Feb 1787	OE	(London, 1956)	arr. of orch version (K 11)	26
2	—	Quartetti: La vera constanza	~1799 [?c1790]	E	—	16 pieces arr. from op (E 19) ? by Haydn or with his approval	
3	—	Quartetti: Armida	~1799 [?c1790]	E	—	18 pieces arr. from op (E 23) ? by Haydn or with his approval	

Note: arrs. of other Haydn op and orat for str qt or qnt, not authenticated; VI Fugen … von G.J. Werner … herausgegeben von … J. Haydn (Vienna, 1804) not arr. but only issued by Haydn

Appendix O.2: Selected spurious works

No.	hIII	Op.	Title, key	Date	Edition	Remarks
1–6			6 Quatuors:			
1	13	3/1	E	~1777	(London, n.d.)	? by R. Hoffstetter, though in HV
2	14	3/2	C			
3	15	3/3	G			
4	16	3/4	B♭			
5	17	3/5	F			II: 'Serenade'
6	18	3/6	A			
7	D 3	—	Divertimento, D	~1763	(Mainz, 1955)	by Albrechtsberger
8	E 2	—	E	~1768	HM, xcviii (1936)	

Note: further spurious qts listed in hIII, and Feder (H1974)

Appendix O.3: Spurious arrangements

No.	H	Op.	Key	Date	Edition	Remarks
1	III:5	1/5	Bb	-1770/71	(London, n.d.)	arr. of sym. J 107; spurious though in HV
2	III:9	2/3	Eb	-1766	(London, n.d.)	arr. of N 11; spurious though in HV
3	III:11	2/5	D	-1766	(London, n.d.)	arr. of N 12; spurious though in HV

50

P: STRING TRIOS (DIVERTIMENTOS)

baryton trios excepted; for 2 violins and cello (or other bass instrument) unless otherwise stated

HV	Key	Date	Authentication	Edition	Remarks
1	E	-1767	EK	HW xi/1, 1; D 901 (1982)	
2	F	-1767	EK	HW xi/1, 9; D 902 (1982)	
3	b	-1767	EK	HW xi/1, 17; D 903 (1985)	
4	Eb	-1767	EK	HW xi/1, 24; D 904 (1981)	
5	B	-?1765	EK	—	lost; forces unknown
6	Eb	-?1764 [-?1761]	EK	HW xi/1, 33; D 905 (1985)	various versions extant, with scherzo as I or III or missing
7	A	-?1765	EK	HW xi/1, 40; D 907 (1982)	incl. variations on Ich liebe, du liebest (G 48); cf appx N 4
8	Bb	-1765	EK	HW xi/1, 45; D 908 (1982)	for vn, va, b
9	Eb	-?1765	EK	—	lost; forces unknown
10	F	-1767	EK	HW xi/1, 55; D 910 (1982)	
11	Eb	-1763	EK	HW xi/1, 68; D 923 (1981)	
12	E	-1767	EK	HW xi/1, 73; D 911 (1981)	
13	Bb	-?1765	EK	HW xi/1, 85; D 912 (1984)	
14	b	-?1765	EK	—	lost; forces unknown
15	D	-1762	EK	HW xi/1, 96; D 914 (1981)	cf appx N 3
16	C	-1765 [-?1763]	EK	HW xi/1, 105; D 915 (1981)	
17	Eb	-1766 [-?1763]	EK	HW xi/1, 117; D 916 (1982)	
18	Bb	-1765 [-?1763]	EK	HW xi/1, 127; D 917 (1984)	
19	E	-1765 [-?1763]	EK	HW xi/1, 135; D 918 (1982)	
20	G	-1766 [-?1763]	EK	HW xi/1, 147; D 919 (1982)	
21	D	?1765	EK	HW xi/1, 157; D 922 (1982)	

Appendix P: Selected works for 2 violins and cello (or other bass instrument) attributed to Haydn (early works if authentic)

No.	HV	Key	Edition	Remarks
1	D3	D	HW xi/2, 29; D 920 (1981)	?authentic; nos. 1, 3, 4 considered doubtful in HW xi/2
2	F1	F	HW xi/2, 6; D 928 (1981)	?authentic

No.	HV	Key	Edition	Remarks
3	G1	G	HW xi/2, 34; D 921 (1981)	?authentic; see remark for no.1
4	A2	A	HW xi/2, 16; D 934 (1985)	?authentic; see remark for no.1
5	A3	A	—	?authentic; nos.5, 8, 9 considered not authentic in HW xi/2
6	D1	D	HW xi/2, 1; D 924 (1981)	?authentic
7	B1	Bb	HW xi/2, 11; D 927 (1981)	?authentic
8	G3	G	(Mainz, 1955); D 933 (1985)	I and II ?authentic; see remark for no.5
9	G4	G	D 926 (1984)	?authentic; see remark for no.5
10	C3	C	D 925 (1985)	doubtful; nos.10–14 considered authentic by Landon (A1980), nos.10, 12–14 as possibly authentic by Larsen (B1941), nos.10, 11, 14, 15 as not authentic by Fruehwald (H1988) and in HW xi/2
11	C2	C	D 931 (1982)	doubtful; see previous remark
12	C1	C	HW xi/2, 40; D 932 (1985)	doubtful; see remark for no.10; nos.12, 13 considered authentic by Fruehwald (1988)
13	C4	C	HW xi/2, 24; D 929 (1982)	doubtful; see remarks for nos.10, 12
14	C5	C	D 930 (1985)	doubtful; see remark for no.10
15	Es4	Eb	—	doubtful; see remark for no.10
16	C6	C	—	probably by Fils (trio for hpd, vn, b)
17	C7	C	5 Eisenstädter Trios (Wiesbaden, 1954), no.3	probably not authentic
18	C8	C	—	probably not authentic
19	D4	D	6 Weinzierler Trios (Wolfenbüttel, 1938), no.5	probably not authentic
20	—	d	—	A-Wst; probably not authentic; incipit in HW xi/2, preface
21	Es2	Eb	—	probably not authentic
22	Es3	Eb	—	probably not authentic
23	Es5	Eb	12 Menuette (Wolfenbüttel, 1938), nos.9, 12	probaby not authentic; 5 movts; only 2 minuets pubd
24	Es11	Eb	(?Paris, n.d.)	probably not authentic
25	E2	E	—	probably not authentic
26	F7	F	as no.24	probably by Pugnani
27	—	F	—	CZ-Bm (2 vn, vlc); probably not authentic
28	G5	G	as no.24	probably not authentic
29	—	A	—	Pmm; probably not authentic; incipit in HW xi/2, preface
30	B3	Bb	—	probably not authentic; ? by Zappa
31	B4	Bb	—	probably not authentic
32	Es1	Eb	(Munich-Gräfelfing, 1969)	by M. Haydn; for vn, va, vc
33	G2	G	(Leipzig, 1932) (2 vn, pf, vc ad lib)	by M. Haydn; edn as sonata op.8, no.5
34	A1	A	(Leipzig, 1932) (2 vn, pf, vc ad lib)	by M. Haydn; edn as sonata op.8, no.6
35	Es9	Eb	6 Weinzierler Trios (Wolfenbüttel, 1938), no.6	probably by L. Hofmann

Note: other works in HV, probably by other composers, incl. D2, E1, B2: by M. Haydn; C9 (lost), D5, E8, F5 (lost, qt), G6, A7: ? by L. Hofmann; Es12, F2, F6: ? by Kammel; Es10: ? by Asplmayr; E3: ? by Ivanschiz; F3: ? by Asplmayr/Ivanschiz; Es7: ? by P. Gasparini; Es13 (lost): ? by Auffmann; F4: ? by J.C. Bach (qt); A5: ? by Enderle, A6: ? by Fils; B5: ? by Chiesa; D6 is arr. of baryton trio Q 1; incipits of 6 doubtful/spurious works: HV: Es6, F8, G7, A4, B6, and preface to HW xi/2: JHI-C10; see also pf sonatas, W 40–42

Q: BARYTON TRIOS (DIVERTIMENTOS)

for baryton, viola and cello (or other bass instrument); WT = 6 leichte Wiener Trios, Wolfenbüttel, 1939

No.	HXI	Key	Date	Authentication	Edition	Remarks
1–24: Book i						
1	1	A	–14 Jan 1767 [c1765–?]	EK	HW xiv/1, 1	bound by that date; preserved singly [15]; H with II and III reversed and without IV; see also note to appx P
2a	—	?1st version, A		I: EK	HW xiv/1, 6	3 movts
2b	2	2nd version, A		I: EK; III, IV: A (u)	HW xiv/1, 6	4 movts; facs. of autograph frag. in Unverricht (N1969)
2c	2bis	spurious version, G		—	WT, no.4	3 movts as 2a but in order II, III, I; for vc, va, b, and other arrs.; cf appx X.3, 9a–c
3	3	A	–1770		HW xiv/1, 16	
4	4	A		I: EK	HW xiv/1, 21	
5a	—	?1st version, A		I: EK	HW xiv/1, 24	3 movts; I quotes Gluck: Che farò senza Euridice
5b	—	?2nd version, A		I: EK; III: A (u)	HW xiv/1, 24	I, II as in 5a; new minuet and trio as III; facs. of autograph frag. in Unverricht (N1969)
5c	i, 596 (below)	spurious version of 5a, G		—	WT, no.2	for 2 vn, b; also in D. arr. fl, vn, b
5d	5	inc. reconstruction of 5b, A		I: EK; III: A (u)	HW xiv/1, 24	2 movts: I, III of 5b
	6	A	–1769		HW xiv/1, 28	spurious version omits II and incl. III from 5b
	7	A	–1769		HW xiv/1, 34	
	8	A			HW xiv/1, 40	
	9	A	–1770		HW xiv/1, 46	
	10	A	–1772	A (frag., u)	HW xiv/1, 51	autograph not for 2 barytons, b, as stated in H
	11	D	–1772		HW xiv/1, 56	
	12	A			HW xiv/1, 61	
	13	A			HW xiv/1, 70	only I extant or identified; edn in Bb for 2 vn, vc; cf appx Q 1
	14	D			HW xiv/1, 72	
	15	A	–1772		HW xiv/1, 78	
	16	D	–1772		HW xiv/1, 84	
	17	A	–1772		HW xiv/1, 88	
	18	A			—	lost or unidentified
	19	D			HW xiv/1, 96	
	20	D			HW xiv/1, 102	
	21	A	–1771		HW xiv/1, 108	only I extant or identified; edn for 2 vn, vc
	22	A			HW xiv/1, 113	
	23	D			—	lost or unidentified; cf appx Q 2, 3
	24	D	1766	A (inc.)	HW xiv/1, 115	edn with Trio of Minuet and III, both missing in H

No.	HXI	Key	Date	Authentication	Edition	Remarks
25–48: Book ii						
25		A	-11 Oct 1767 [c1766/7]	EK	HW xiv/2, 1	bound by that date
26		G	-1772		HW xiv/2, 6	uses minuet of appx X.1, 2
27		D			HW xiv/2, 13	II and III reversed in H
28		D			HW xiv/2, 19	
29		A			HW xiv/2, 25	I uses theme from La canterina (E 8)
30		G			HW xiv/2, 32	
31		D			HW xiv/2, 37	another version (?not authentic) has 4 movrs, incl. Adagio from no.5
32		G			HW xiv/2, 43	
33		A			HW xiv/2, 49	
34		D	-1776 [-?1775]		HW xiv/2, 56	
35		A	-1771		HW xiv/2, 61	
36		D	-1776		HW xiv/2, 65	
37		G	-1776		HW xiv/2, 70	'I uses W 3
38		A	-1776		HW xiv/2, 77	
39		D	-1776		HW xiv/2, 83	
40		D		A (frag.)	HW xiv/2, 88	
41		D		A	HW xiv/2, 93	
42		D	1767	A (frag.)	HW xiv/2, 98	
43		D			HW xiv/2, 104	
44		D			HW xiv/2, 109	
45		D			HW xiv/2, 114	
46		A			HW xiv/2, 120	
47		G			HW xiv/2, 125	
48		D			HW xiv/2, 131	
49–72: Book iii						
49		G	-7 July 1768 [c1767–8]	EK	HW xiv/3, 1	bound by that date
50		D			HW xiv/3, 7	
51		A			HW xiv/3, 14	
52		d/D			HW xiv/3, 18	minuet and trio based on movr in sym. J 58
53		G	1767	A	HW xiv/3, 24	
54		D			HW xiv/3, 29	
55		G			HW xiv/3, 33	
56		D			HW xiv/3, 38	
57		A	1768	A	HW xiv/3, 44	
58		D			HW xiv/3, 48	
59		G		Sk	HW xiv/3, 53	

No.	hXI	Key	Date	Authentication	Edition	Remarks
	60	A		Sk	HW xiv/3, 59	for sketch, see critical commentary to HW xiv/3, 32
	61	D		Sk	HW xiv/3, 65	for sketch, see critical commentary to HW xiv/3, 35
	62	G			HW xiv/3, 72	
	63	D			HW xiv/3, 77	
	64	D			HW xiv/3, 83	I uses Alleluia theme of sym. J 30
	65	G			HW xiv/3, 88	
	66	A			HW xiv/3, 93	
	67	G			HW xiv/3, 100	
	68	A		A	HW xiv/3, 106	
	69	D		A	HW xiv/3, 111	
	70	G			HW xiv/3, 116	
	71	A			HW xiv/3, 122	
	72	D			HW xiv/3, 128	paper for copying ordered by that date; bound by 3 Feb 1772
73–96: Book iv						
	73	G	−22 Dec 1771 [c1768–71]	EK, SC	HW xiv/4, 1	
	74	D	−1772		HW xiv/4, 5	
	75	A			HW xiv/4, 11	
	76	C			HW xiv/4, 16	
	77	G	−1772		HW xiv/4, 21	
	78	D			HW xiv/4, 26	
	79	D	1769	A	HW xiv/4, 30	
	80	G		A (frag.)	HW xiv/4, 35	
	81	D			HW xiv/4, 41	
	82	C			HW xiv/4, 46	
	83	F			HW xiv/4, 52	
	84	G			HW xiv/4, 58	
	85	D			HW xiv/4, 64	
	86	A			HW xiv/4, 70	
	87	a			HW xiv/4, 76	
	88	A			HW xiv/4, 82	
	89	G			HW xiv/4, 87	vn instead of va
	90	C			HW xiv/4, 93	vn instead of va
	91	D			HW xiv/4, 100	vn instead of va
	92	G			HW xiv/4, 106	
	93	C			HW xiv/4, 111	
	94	A	−1774		HW xiv/4, 116	
	95	D			HW xiv/4, 123	
	96	b			HW xiv/4, 130	

No.	HXI	Key	Date	Authentication	Edition	Remarks
97–126*: Book v						
97	D		−8 Nov 1778 [c1771–8]	EK	HW xiv/5, 1	paper for copying ordered by that date; preserved singly
98	D		[–c1773] [?1766]	EK	HW xiv/5, 15	'per la felicissima nascita di S.Al.S. Prencipe Estorházi'; cf S 11
99	G			EK	—	lost
100	F			EK	HW xiv/5, 22	
101	C		[–c1773]	EK	HW xiv/5, 30	
102	G			EK	HW xiv/5, 37	
103	A			EK	HW xiv/5, 44	I and probably II based on U 15; see also V 6
104	D			EK	HW xiv/1, 126	MS discovered 1976, see Fisher (B1978)
105	G		1772	A	HW xiv/5, 50; xiv/1, 132	III discovered 1976, see Fisher (B1978)
106	D		[–c1773]	EK	HW xiv/5, 55	autograph erroneously mentioned in H is that of no.105
107	D		[?c1766–8]	EK	HW xiv/5, 61	
108	A			EK	HW xiv/5, 68	
109	C		[–c1773]	A, EK	HW xiv/5, 74	
110	C		[–c1773]	EK	HW xiv/5, 80	I and II based on U 13
111	G			EK	HW xiv/5, 87	in HV 'a cinque'
112	D			EK	HW xiv/5, 92	
113	D		[–c1773]	HV, JE	HW xiv/5, 99	
114	D		[–c1773]	EK	HW xiv/5, 106	
115	G			EK	HW xiv/5, 113	
116	F			EK	HW xiv/5, 119	
117	D		[–c1773]	EK	HW xiv/5, 125	
118	G			EK	HW xiv/5, 132	
119	D			EK	HW xiv/5, 139	only baryton pt extant
120	A		[–c1773]	EK	HW xiv/5, 141	
121	A			HV, C	HW xiv/5, 146	
122	G			EK	HW xiv/5, 153	
123	G			EK	HW xiv/5, 159	
124	G			EK	HW xiv/5, 166	
125	G			EK	HW xiv/5, 174	
126*	C			EK	HW xiv/5, 180	

Appendix Q: String (probably baryton) trios attributed to Haydn

No.	HXI	Title, key	Forces	Edition	Remarks
1	D1 (I, II)	Adagio cantabile, D; Menuetto (with Trio), A/a	2 vn, vc	HW xiv/1, 120, 122	probably authentic; ? II, III of Q 13
2	iii, 327	Finale (Presto assai), D	2 vn, vc	HW xiv/1, 124	probably authentic; ? III of Q 23

No.	HXI	Title, key	Force	Edition	Remarks
3	iii, 327	Menuetto (with Trio), Eb	2 vn, vc	HW xiv/1, 123	probably authentic, but transposed; ? II of Q 23
4	D2	Divertimento, D	baryton, va, vc	—	doubtful
5	A1	Terzetto (a tre), A	baryton, va b	—	doubtful
6	C3	Trio, C	vc, va, b	DTÖ, cxxiv (1972), 71, as by Tomasini	baryton trio by Luigi Tomasini

Note: for HXI:C1–2, see appx S 27, 26; see also appx S 1

R: WORKS FOR 1 OR 2 BARYTONS

No.	H	Title, key	Force	Date	Authentication	Edition	Remarks
1–5		[5] Divertimenti per il pariton solo:					
1	XII:20	G	?with vc	c.765/6	EK	—	lost
2	XII:21	D	?with vc	c.765/6	EK	—	lost
3	XII:22	A	?with vc	c.765/6	EK	—	lost
4	XII:23	G	?with vc	c.765/6	EK	—	lost
5	XII:18	A	with vc	c.765–9	EK	—	lost
6–7		[2] Soli per il pariton:					
6	XII:13	D	with vc	?1770–75	EK	—	lost; probably based on U 15; theme identical with III of V 6
7	XII:14	D	with vc	?1770–75	EK	—	lost
8–13		6 Sonate:	baryton, vc	?1775			
8	XII:7	D			EK	—	lost
9	XII:8	C			EK	—	lost
10	XII:9	G			EK	—	lost
11	XII:10	A			EK	—	lost
12	XII:11	D			EK	—	lost
13	XII:12	G			HV	—	lost
14–16		[3] Sonate:	baryton, vc	?	HV	—	lost; on authenticity see critical commentary to HW xiii, 11; ?incl. in '16 Duetten für den Bariton' in Haydn's short work-list, c1803/4 (H i, facs., p.V)
17–22		[6] Duetti:	? barytons	c1764–9			
17	X:11	D			EK	HW xiii, 2	only extant in unauthentic arr. for fl, vn, b
18	XII:16	G			EK, JE	HW xiii, 5	
19	XII:1	A			EK	HW xiii, 10	as no.17

No.	H	Title, key	Date	Authentication	Edition	Remarks
20	XII:5 + 3	D		EK	HW xiii, 16	as no.17; finale uses Trio from sym. J 28
21	XII:6	G		EK	—	lost
22	XII:2	G		HV	—	lost
23	XII:19	12 Cassations-Stücke (Divertimentos) 2 barytons, b	c1765/6	EK, A (small frag., u); JE	HW xiii, 20	
24	X:9	Divertimento, D 2 barytons, 2 hn	?1765–70	EK	—	lost
25–6		[2] Divertimentos (Quintetti): 2 hn, baryton, va, b				'3 Quintetten' according to Haydn's short work-list, c1803/4 (H i, facs., p.V)
25	X:7	D	c1767/8	EK	—	lost
26	X:10	D	c1767/8 [?c1771]	EK	HW xiii, 29	uses N 32
27–33		[7] Divertimentos a 8: 2 hn, 2 vn, baryton, va, vc, vle				1st edn with fl instead of baryton 22, 70
27	X:2	D	?1775	EK	HW xiii, 38	only extant in arrs. without baryton
28	X:5	G	1775	EK, A	HW xiii, 62	
29	X:3	a/A	1775	EK, A	HW xiii, 87	
30	X:4	G	?1775	EK	HW xiii, 109	as for no.27
31	X:1	D	1775	EK, A	HW xiii, 131	autograph in PL-Kj
32	X:6	A	?1775	EK	HW xiii, 157	as for no.27
33	X:12*	G	?1775	—	HW xiii, 177	as for no.27

Note: HX:8 probably = XI:111 (Q 111); XII:24 probably = XI:114 (Q 114); XII:25* has not been verified

S: MISCELLANEOUS CHAMBER MUSIC FOR 2–3 STRING AND/OR WIND INSTRUMENTS

No.	H	Title, key	Forces	Date	Authentication	Edition	Remarks
1–6		6 Violin Solo mit Begleitung einer Viola (Sonatas; Duos):	vn, va	−1773 [−?1769]	EK	(Mainz, 1970)	'1773' on MS of nos.1, 4 in A-Wn; versions for 2 vn and vn, vc, doubtful 70
1	VI:1	F		A (u vn)			
2	VI:2	A		A (u vn)			cf. appx L 11
3	VI:3	Bb					
4	VI:4	D					
5	VI:5	Eb					cf. appx L 11
6	VI:6	C					
7	IV:5*	Divertimento a 3 per il corno di caccia, Eb	hn, vn, vc	1767	A	D 1 (1957)	
8–13		6 Divertimentos a 3 (Divertissements): vn/fl, vn, vc		1784	SC	D 871–6 (1989), ed. E. Kubitschek	op.100 in early edns 70

No.	H	Title, key	Forces	Date	Authentication	Edition	Remarks
8	IV:6	D					I arr. from Il mondo della luna, no.12 (E 17); II uses no.15
9	IV:7	G					I uses no 24 of E 17
10	IV:8	C					II uses no.25 of E 17
11	IV:9	G					ar. of 3 movts from Q 97
12	IV:10	A					II arr. from no.23 of E 17
13	IV:11	D					III uses no.14 of E 17
14–17		[?4] Trios:	2 fl, vc	1794/5			orig. versions of nos.15–17 uncertain 70
14	IV:1	C		1794	A	(Leipzig, 1959), ed. K.H. Köhler	2, not 3 versions of II, both in edn, but 2nd version spoiled
15	IV:2	G			A (u)	as no.14	1 movt only; II only final variation of I; in autograph without title or 'Fine' remark, contrary to statement in H; uses song The Lady's Looking-Glass (appx G.1, 2); 1st pubd with added III from no.16 (London, 1799) autograph in PL-Kj
16	IV:3	G (no.2)			A (u)	as no.14	
17	IV:4	G (no.3)			E	NM 71 (1954)	1 movt only; MS title in Haydn's hand, not signed by him, no author's name; also with no.15 added as II

Appendix S: Works attributed to Haydn

No.	H	Title, key	Forces	Date	Authentication	Edition	Remarks
1	IV:G2	Divertimento a 3, G	fl, vn, vc	?1762		—	?authentic; 1 movt only; ?arr. of otherwise lost baryton composition; date on MS copy
2	VI:C4	Violino solo (Arioso + 7/8 variations), C	vn, b	–1768			doubtful
3–5	VI, Anhang	3 movrs (each 2nd trio of Minuet) i F; ii 2 vn Bb; iii c		i 1800 ii–iii –1802		i (Paris, 1800); ii–iii (Leipzig, 1917)	doubtful; ?arr. from unknown works
6	VI:G2	Solo con basso, G	vn, b	?			doubtful
7	VI:Es2 Add.	23 variations, Eb	vn, b	?		ed. A. Weinmann (Vienna, 1982)	doubtful; MS copy, A-SEI
8	IV:D2	Cassation, D	fl, vn, b	?		(Frankfurt, 1973) (with added hpd)	doubtful; arr. for fl, vn, str, hpd (Frankfurt, 1973)
9	IV:D1	Trio, D	fl, vn, b	–1768			doubtful
10–12		3 Terzetti:	fl/vn, va, vc/bn	?		—	probably not authentic; MS copy, HE
10	—	C					
11	—	F					

119

No.	H	Title, key	Forces	Date	Edition	Remarks
12	—	B♭				
13	IV:F1	Divertimento (Trio), F	3 fl	?	—	probably not authentic
14	IV:F2	Sonata a 3, F	lute, vn, b	?	(Antwerp, 1973)	probably not authentic
15–17		3 Trios	clarinetto d'amour, vn, b	–1781	(Leipzig, 1977)	probably not authentic
15	IV:Es1	E♭				
16	IV:Es2	E♭				
17	IV:B1	B♭				
18	VI:G4	Ein musikalischer Scherz, G	2 vn	?	(Offenbach, 1896)	probably not authentic
19	IV:D3	Divertimento, D	hn, va, vle	?	ed. W. Rainer, D 274 (1969), as by M. Haydn	? by M. Haydn
20–23		4 Duos:	vn, va	1783	(Leipzig, 1911), as by M. Haydn	by M. Haydn
20	VI:C1	C				
21	VI:E1	E				
22	VI:F2	F				
23	VI:D3	D				
24–5		[2] Divertimenti da camera:		?	(Wolfenbüttel, 1972)	probably by Haver (? Gregor Hauer)
24	IV:G1	G	vn/fl, vn, b			also attrib. Vanhal
25	IV:A1	A	fl, vn, b			
26	XI:C2	Divertimento, C	fl, vn, b	?	(London, 1851)	probably by Haver
27	XI:C1	Divertissement, C	vn/fl, vn, b	–1771	(London, 1936) (for vn, pf)	probably by Dittersdorf (2 vn, b)
28	VI:D1	Duett, D	vn, vc	–1768	(Wiesbaden, 1982)	probably by L. Hofmann
29–34		6 Sonatas:	2 vn	–1770	(Mainz, 1953)	also attrib. Campioni; probably by Kammel
29	VI:G1	G				
30	VI:A1	A				
31	VI:B1	B♭				
32	VI:D2	D				
33	VI:Es1	E♭				
34	VI:F1	F				
35	IV, Anh.	Gioco filarmonico, D	2 vn/fl, b	–1781	(Naples, –1790)	by M. Stadler; also pubd for pf
36	i, 509	Divertimento, E♭	vn, va d'amore, vc	?	NM 52 (1930)	spurious arr. of Q 56, I; Q 34, II (trio of minuet by Gassmann); Q 78, III

Note: 6 vn duettos mentioned by A. Fuchs, doubtful and lost (hVI:G3, D4, A2, C2, F3, C3); arr. for lute/gui, vn, vc, see O 1, 6; 3 ob duettos, no.1 on Teldec 6.42416 AW, doubtful

T: WORKS FOR 2 LIRE ORGANIZZATE

No.	H	Title, key	Forces	Date	Authentication	Edition	Remarks	
1–5		[?5] Concerti per la lira organizzata:	2 lire, 2 hn, 2 vn, 2 va, vc				1 conc., ? in C, possibly lost; nos.1–3 ?1st ser., nos.4–5 ?rest of 2nd ser.	26, 66
1	VIIh:1	C		?1786	HE (= C)	HW vi, 1		
2	VIIh:4	F		[1786]	HE	HW vi, 35		
3	VIIh:2	G		?1786	HC	HW vi, 75	only extant MS copy without author's name but rev. Haydn; II uses cavatina Sono Alcina (≈8)	
4	VIIh:5	F		?1787	HC	HW vi, 113	only extant MS copy without a uthor's name but rev. Haydn; cf J 89	
5	VIIh:3	G		?1787	A (u frag.). HC	HW vi, 141	only extant MS without author's name but partly written by Haydn; cf J 100	66
6–13		[?8] Notturni:					?1 notturno missing: nos.6–11	26, 32, 70
6	II:25	C	2 lire, 2 cl, 2 hn, 2 va, b	c1788–90	HE (= C)	HW vii, 1	?1st ser., nos.12–13 ?rest of 2nd ser.	
7	II:26	F	2 lire, 2 cl, 2 hn, 2 va, b	c1788–90	HC	HW vii, 25	only extant MS copy without author's name but from Haydn's collection	
8	II:32	C, i orig. version	2 lire, 2 cl, 2 hn, 2 va, b	?1790	RC	HW vii, 48 (based on both versions)	MS copy without author's name but rev. Haydn	
		ii London version	2 fl, 2 vn, 2 hn, 2 va, vc, db	?1792	RC		MS copy without Haydn's name but rev. Haydn	
9	II:31	(Divertimento), C, i orig. version	2 lire, 2 cl, 2 hn, 2 va, vc	1790	A, Sk	HW vii, 78	sketch for I, HW vii, 188	
		ii 1st London version	fl, ob, 2 cl/vn, 2 hn, 2 va, vc	?1792	SC, RC	HW vii, 78		
		iii 2nd London version	fl, ob, 2 hn, 2 va, vc, db	?1794	corrections in A	HW vii, 78		
10	II:29	C, i ?orig. version	? 2 lire, ? 2 cl, ? 2 hn, ? 2 va, ? vc	?1790	—	[HW vii, 98]		
		ii extant version	fl, ob, 2 vn, 2 hn, 2 va, vc/b	?1791	HE	HW vii, 98		
11	II:30	G	2 lire, 2 cl, 2 hn, 2 va, vc	?1790	HC, C	HW vii, 116		
12	II:28	F, i ?orig. version	? 2 lire, ? 2 hn, ? 2 vn, ? 2 va, vc, ? vc	?1790	Sk	[HW vii, 130]		
		ii London version	fl, ob, 2 hn, 2 va, 2 va, vc, db	?1792	SC	HW vii, 130		
13	II:27	(Divertimento), G, i orig. version	2 lire, 2 hn, 2 vn, 2 va, vc	?1790	A	HW vii, 158 (based on both versions)	finale lost; only extant MS copy without author's name	
		ii London version	fl, ob, 2 hn, 2 vn, 2 va, vc, db	?1792	corrections in A, SC	HW vii, 196	sketch for I, HW vii, 196	

121

U: KEYBOARD CONCERTOS/CONCERTINOS/DIVERTIMENTOS

No.	H	Title, key	Forces	Date	Authentication	Edition	Remarks	
1–2		[2] Concerti per il clavicembalo:						
1	XVIII:1	(Concerto per l'organo, no.1), C	org/hpd, 2 ob, (2 tpt/?hn, ? timp), str	?1756	A, EK	(Wiesbaden, 1953, 2/1986)	2nd edn with tpts	9, 11, 67
2	XVIII:2	D	org/hpd (2 ob, 2 tpt, timp), str	–1767	EK	D 78 (1997)	attrib. Galuppi in MS copy, *D-Bsb*	
3	XVIII:6	Concerto per violino e cembalo, F	org/hpd, vn solo, str	1766	EK	(Kassel, 1959)		67
4–6		[3] Concerti per il clavicembalo:						
4	XVIII:3	F	hpd, (?2 hn), str	1771 [–?c1766]	EK	HW xv/2, 1		67
5	XVIII:4	G	hpd/pf, (?2 ob, ?2 hn), str	1781 [?c1768–70]	EK	HW xv/2, 45		
6	XVIII:11	D	hpd/pf, 2 ob, 2 hn, str	1784	—	HW xv/2, 79		
7–13		[7] Concertinos/Divertimentos:						
7	XIV:11	Concertino, C	hpd, 2 vn, b	1760	A (lost)	HW xvi, 1; D 21 (1959)		10, 67
8	XIV:10	Divertimento no.1 con violini, C	hpd, (2) vn, (b)	?c1764–7	JE	HW xvi, 73	only hpd extant; finale uses that of kbd sonata, appx W.1, *11*; facs. in Landon (A1980), 546	67
9	XIV:4	Divertimento (Concerto), C	hpd, 2 vn, b	1764	A	HW xvi, 51		67
10	XIV:3	Divertimento (Concertino, Sonate), C	hpd, 2 vn, b	1771 [–c1767]	EK	HW xvi, 66		67
11	XIV:7	Divertimento, C	hpd, 2 vn, vc	–c1767	HE	HW xvi, 75		67
12	XIV:9	Divertimento, F	hpd, 2 vn, vc	–c1767	HE	HW xvi, 84		67
13	XIV:8	Divertimento, C	hpd, 2 vn, vc	c1768–72	HE	HW xvi, 92	cf Q *110*	67
14	XIV:1	Divertimento, E♭	hpd, 2 hn, vn, b	1766	EK	HW xviii/1, 157	lost; version as pf trio (V 6) extant; cf Q *103*, R *6*	
15	XIV:2	Divertimento, F	hpd, 2 vn, baryton	?c1767–71	EK	—		

Appendix U: Selected works attributed to Haydn

No.	H	Title, key	Forces	Date	Edition	Remarks	
1	XVIII:5	Concerto, C	org/hpd (?2 ob, ?2 tpt/hn), 2 vn, b	–1763	NM 200 (1959)	probably authentic; ?EK; also attrib. Wagenseil in MS copy, *A-El*	
2	XVIII:8	Concerto (no.2), C	org/hpd, (2 hn/tpt, timp), 2 vn, b	–1766	D 80 (1962)	probably authentic; ?EK; orig. attrib. (L.) Hofmann in MS copy, *D-Bsb*	
3	XVIII:10	Concertino (Concerto), C	org/hpd, 2 vn, b	–1771	(Munich, 1969)	probably authentic; ?EK; only extant MS copy, *A-Wgm*, as by 'Heyden'	
4	XIV:12	Concerto (Partita, Concertino), C	hpd, 2 vn, b	–1772 [–c1767]	HW xvi, 10; D 323 (1969)	probably authentic	67

No.	H	Title, key	Forces	Date	Edition	Remarks	
5	XIV:13	Concerto (Concertino), G	hpd, 2 vn, b	~c1767	HW xvi, 26, (Mainz, 1956)	probably authentic; date 1765 not in Görtweig catalogue, contrary to preface in Mainz edn	67
6	XVIII:F2	Concerto (Concertino), F	hpd, 2 vn, b	~c1767	HW xvi, 38, D 324 (1969)	probably authentic	67
7	XIV:C2	Divertimento, C	hpd, 2 vn, b	~c1767	HW xvi, 111; D 325 (1969)	I and II probably authentic; III in HW xvi, 114, doubtful	
8	XIV:C1	Divertimento, C	hpd, (?2) vn, b	~1772 [~1767]	HW xvi, 105, D 534 (1976)	?authentic; allegedly not approved by Haydn in 1803; ? vn 1 lost; in edns as pf trio; ? str spurious	
9	XVIII:G2	Concerto duetto, G	2 hpd/pf, 2 hn, 2 vn, b	~1782	(London, 1782)	by J.A. Steffan (Šetková no.135)	
10	XVIII:7	Concerto, F	org/hpd, 2 vn b	~1766	(Amsterdam, 1962), (Vienna, 1983)	doubtful; considered probably authentic by Larsen (B1941); I and III later versions of pf trio, appx V.1, 8; orig. attrib. Wagenseil in MS copy, CS-KRa	
11	XIV:G1	Partita (Divertimento), G	hpd, 2 vn, b	~1774	—	lost, doubtful; allegedly not approved by Haydn in 1803; not with 2 bn as in H	
12	XVIII:9	Concerto, G	hpd, 2 vn, b	~1767	(Mainz, 1967?), HW xv/2, 131	doubtful; considered probably authentic by Larsen (B1941), spurious by Fruehwald (H1988)	
13	XVIII:Es1	Concerto, Eb	hpd, str	?	—	probably not authentic; orig. without author's name in only extant MS copy, D-Bsb	
14	XVIII:G1	Concerto, G	hpd, 2 fl/ob, 2 hn, 2 vn, b	?	—	probably not authentic	
15	—	Concertino, D	hpd, 2 vn, vc	?	—	[XIV:?]D1 in Brown (O1986); probably not authentic	
16	XVIII:F1	Concerto, F	hpd, 2 fl, str	c1779/80	(Berlin, 1927)	by G.J. Vogler (6 leichte Clavierconcerte, no.6)	
17	XVIII:F3	Concerto, F	hpd, str	~1766	—	probably by J.G.) Lang	
18	—	Concerto (Kleines Konzert), F	hpd, (2 hn), str	~1775	(Heidelberg, 1962)	[XVIII:]F4 in Brown (O1986); by L. Hofmann	
19	XIV:C3	Concerto (Quattro), C	hpd, 2 vn, b	~1766	(Paris, c1776), as by Wagenseil	by Wagenseil	
20	XIV:Es1	Divertimento, Eb	hpd, 2 vn, b	?	—	by J.A. Steffan (orig. for hpd solo)	
21	XIV:F1	Quartetto concertant, F	hpd/harp, fl, vn/va, vc	~1774	(Paris, 1777)	by J. Schmittbauer; orig. for hpd/pf, fl, vn, vc	
22	XIV:F2	Concertante, F	pf, ob, vn, va, vc	~1782	(London, ?c1785), as by J.C. Bach	not a Haydn autograph; by J.C. Bach	

V: KEYBOARD TRIOS

No.	HXV	Title, key	Forces	Date	Authentication	Edition	Remarks	
1	5	Sonata, G	hpd, vn, vc	~25 Oct 1784	A (frag.), SC	HW xvii/2, 1; D 502 (1976)	no.3 of 3 Sonatas, nos.1 and 2 spurious; see appx V.2, 1–2	24, 25
2–4		3 Sonatas:	hpd, vn, vc			D 503–5 (1975–6)		
2	6	F		1784	A (lost frag.), OE	HW xvii/2, 22		25
3	7	D		1785	A (incl. Sk)	HW xvii/2, 39	sketch of III, HW xvii/2, 260	25

No.	HXV	Title, key	Forces	Date	Authentication	Edition	Remarks	
4	8	Bb	hpd, vn, vc	–26 Nov 1785	OE	HW xvii/2, 55		25
5–7		3 Sonatas:						
	9	A		1785		HW xvii/2, 73; D 506 (1975)		25
	2	(Divertimento), F	(hpd, vn, b)	?c1767–71	A, EK, SC	HW xvii/1, 141; D 501 (1976)	uses lost divertimento U 15; cf R 6	25
	10	Eb		–28 Oct 1785	SC	HW xvii/2, 88; D 507 (1975)		25
8–10		3 Sonatas:	hpd/pf, vn, vc			D 508–10 (1973–4)		
	11	Eb		–8 March 1789 [–?16 Nov 1788]	OE	HW xvii/2, 106		25
	12	e		–8 March 1789 [1788/9]	OE	HW xvii/2, 124		25
	13	c		[–29 March] 1789	OE	HW xvii/2, 148		25, 53
11	14	Sonata, Ab	hpd/pf, vn, vc	[?–11 Jan] 1790	OE	HW xvii/2, 169; D 511 (1973)		25
12	16	Trio, D	hpd/pf, fl, vc	[–28 June] 1790	OE	HW xvii/2, 195; D 512 (1970)		25
13	15	Trio, G	hpd/pf, fl, vc	[–28 June] 1790	OE	HW xvii/2, 220; D 513 (1970)		25
14	17	Trio, F	hpd/pf, fl/vn, vc	–?20 June 1790	Sk, ?OE	HW xvii/2, 245; D 514 (1970)	sketch for I in Beßb 1973–7	25
15	32	Sonata, G	pf, vn, vc	–14 June 1794	HV, ?Gr, Dies	HW xvii/3, 313; D 481 (1970)	? orig. for pf, vn	36
16–18		3 Sonatas:	pf, vn, vc	–15 Nov 1794	HV, ?Gr, Dies	D 482–4 (1970)		
	18	A				HW xvii/3, 1		
	19	g				HW xvii/3, 24		
	20	Bb				HW xvii/3, 45		
19–21		3 Sonatas:	pf, vn, vc	–23 May 1795	HV, Gr, Dies	D 485–7 (1970)		36
	21	C				HW xvii/3, 64		
	22	Eb			SC of II (pf only)	HW xvii/3, 85	SC shows slightly different earlier version ?1794/5 of II, ?for pf solo and printed thus in HW xvii/3, 339, D 486, p.36	
	23	d		–9 Oct 1795	HV, ?Dies	HW xvii/3, 114		33, 36, 72, 73
22–4		3 Sonatas:	pf, vn, vc	[?1794]	Sk	D 488–90 (1970)		
	24	D				HW xvii/3, 135		
	25	G				HW xvii/3, 152	III: Gypsy rondo (all'ongarese)	66
	26	f♯				HW xvii/3, 170	see sym. J 102	
25–7		3 Sonatas:	pf, vn, vc	–20 April 1797	HV	D 493–5 (1970)		36, 38, 72
	27	C				HW xvii/3, 190		
	28	E				HW xvii/3, 220		
	29	Eb		[–?Aug 1795]		HW xvii/3, 241		

No.	HXV	Title, key	Forces	Date	Authentication	Edition	Remarks	
28	31	Sonata, e♭	pf, vn, vc	1795	A	HW xvii/3, 265; D 491 (1970)	II orig. dated 1794 and called 'Jacob's Dream!' in autograph; ? orig. a separate work	36
29	30	Sonata, E♭	hpd/pf, vn, vc	–7 Oct 1797 [? 16 April–9 Nov 1796]	A (u frag.); E; OE	HW xvii/3, 284; D 492 (1970)		38

Note: Haydn apparently composed no sonatas for pf, vn, except perhaps no.15; extant edns are arrs., especially of O 66–7, V 28, W 22–4, 37 (HXVI:63bis), appx W.2, 1

Appendix V.1: Early trios for harpsichord, violin and bass attributed to Haydn

No.	HXV	Title, key	Date	Authentication	Edition	Remarks	
1	36	Partita (Concerto), E♭	–1774 [–?1760]	H 1803	HW xvii/1, 1; D 530 (1977)	probably authentic	
2	C1	Divertimento, C	–1766 [–?1760]	—	HW xvii/1, 13; D 522 (1977)	not verified by Haydn in 1803; also attrib. Wagenseil (see H. Scholz-Michelitsch: *Des Orchester- und Kammermusikwerk von Georg Christoph Wagenseil: thematischer Katalog* (Vienna, 1972), no 449); early edn with minuet and trio from J.-P.-E. Martini: L'amoureux de quinze ans	
3	37	Divertimento (Trio, Concerto), F	–1766 [–?1760]	H 1803	HW xvii/1, 31; D 521 (1977)	probably authentic	
4	38	Divertimento, B♭	–1769 [–?1760]	H 1803	HW xvii/1, 45; D 531 (1977)	probably authentic	
5	34	Partita (Divertimento), E	–1771 [–?1760]	H 1803, F	HW xvii/1, 57; D 22 (1959), 529 (1977)	probably authentic	
6	f1	Partita, f	–?1760	F	HW xvii/1, 67; D 532 (1977)	probably authentic	50, 71
7	41	Divertimento, G	–1767 [–?1760]	H 1803	HW xvii/1, 81; D 527 (1977)	probably authentic	
8	40	Divertimento (Partita), F	–1766 [?c1760]	—	HW xvii/1, 97; D 526 (1977)	probably authentic; one MS copy with spurious Adagio (see D 4, p.8) instead of minuet; see appx U 10	
9	1	Partita (Divertimento), g	–1766 [?c1760–?2]	H 1803	HW xvii/1, 109; D 525 (1977)	probably authentic	71
10	35	Divertimento (Capriccio), A	–1771 [?c1764/5]	HE, H 1803	HW xvii/1, 123; D 528 (1977)	probably authentic	
11	33	Divertimento, D	–1771 [–?1760]	H 1803	HW xvii/1, 175 (incipits)	lost	
12	D1	Divertimento, D	–1771	—	HW xvii/1, 175 (incipits)	lost; doubtful, not verified by Haydn in 1803; according to Pohl, fcr hpd, 2 vn, vc	71

Appendix V.2: Doubtful and spurious works and arrangements

No.	H	Title, key	Forces	Date	Edition	Remarks	
1	XV:3	Sonata, C	hpd/pf, vn, vc	−1784	HW xvii/2, appx, 261	probably by Pleyel, though mentioned in Haydn's contract with Forster, 1786, and one MS copy signed by Haydn; orig. without vc; see V 1	24
2	XV:4	Sonata, F	hpd/pf, vn, vc	−1784	HW xvii/2, appx, 280	probably by Pleyel; see above remarks	24
3	XV:C2	Grand bataille, C	pf, vn, vc	c1800	(Paris, c1804–14)	arr. from syms. J 48, I, J 76, III, J 81, III, with spurious movts added	
4	XIV:6	Sonata, G	hpd, vn, vc	−1767	D 523 (1977)	arr. of sonata W 1	
5	XV:39	Sonata, F	hpd, vn, vc	−1767	D 524 (1977)	arr. from sonatas appx W.1, 11, 10, 2, with spurious Andante	
6	XV:42 Add.	Variazioni, D	hpd, vn, b	?	D 533 (1976)	apparently arr. from otherwise unknown movt (see appx X.1, 3) and from appx X.1, 4	

Note: see also appx U 8

W: KEYBOARD SONATAS

Editions *Joseph Haydn: Sämtliche Klaviersonaten*, I–III, ed. C. Landon (Vienna, 1964–6) [WU]
Joseph Haydn: Sämtliche Klaviersonaten, I–III, ed. G. Feder (Munich, 1972) [HU], mostly identical with HW xviii/1–3

No.	HXVI	Title, key	Instrument	Date	Authentication	Edition	Remarks	
1	6	Partita (Divertimento), G	hpd	−1766 [−?1760]	A (no IV)	HU i, 34; WU 13	see appx X.3, 10	10
2	14	Parthia (Divertimento), D	hpd	−1767 [−?1760]	EK	HU i, 26; WU 16		
3	3	Divertimento, C	hpd	[?c1765]	EK	HU i, 98; WU 14	see baryton trio Q 37	
4	4	Divertimento, D	hpd	[?c1765]	EK	HU i, 104; WU 9	III and IV in H not part of this work; ? orig. III lost	
5	2a	Divertimento, d	hpd	[?c1765–70]	EK	—	lost; WU 21; 6 sonatas announced in 1993 as 18 rediscovered nos.5–10 are modern forgeries using Haydn's incipits	
6	2b	Divertimento, A	hpd	[?c1765–70]	EK	—	lost; WU 22	
7	2c	Divertimento, B	hpd	[?c1765–70]	EK	—	lost; WU 23	
8	2d	Divertimento, B♭	hpd	[?c1765–70]	EK	—	lost; WU 24	
9	2e	Divertimento, e	hpd	[?c1765–70]	EK	—	lost; WU 25	
10	2g	Divertimento, C	hpd	[?c1765–70]	EK	—	lost; WU 26	
11	2h	Divertimento, A	hpd	[?c1765]	EK	—	lost; WU 27	
12a	47bis Add.	Divertimento, e	hpd		—	HU i, 108; WU 19	earlier and probably orig. version of no.12b	
12b	47	Sonata, F	hpd/pf	−1788	HV	WU 57	doubtful, though apparently authorized version of no.12a; doubtful Moderato added as I, Minuet omitted	

No.	HXVI	Title, key	Instrument	Date	Authentication	Edition	Remarks	
13	45	Divertimento, E♭	hpd	1766	A	HU i, 116; WU 29		18
14	19	Divertimento, D	hpd	1767	A	HU i, 130; WU 30		18
15	5a Add. = XIV:5	Divertimento, D	hpd	[c1767–70]	EK, A (u frag.)	HU i, 143; WU 28	frag.; only I (inc.) and II extant	18
16	46	Divertimento, A♭	hpc	–1788 [c1767–70]	EK	HU i, 147; WU 31		18
17	18	Sonata, B♭	hpc	–1788 [c1771–3]	A (u frag.), HV	HU i, 162; WU 20		19
18	44	Sonata, g	hpc	–1788 [c1771–3]	HV; H 1799	HU i, 171; WU 32		19
19–24		6 Sonatas:	hpc	–Feb 1774	OE, EK			
19	21	C		1773	A (frag.)	HU ii, 1; WU 36		19
20	22	E		1773	A	HU ii, 12; WU 37		19
21	23	F		1773	A (frag.)	HU ii, 22; WU 38		19
22	24	D		?1773		HU ii, 34; WU 39		19
23	25	E♭		?1773		HU ii, 44; WU 40		19
24	26	A		1773	A (no.II)	HU ii, 52; WU 41; facs. (Munich, 1958)	minuet and trio arr. from sym. J 47	19
25–30		6 Sonatas	hpd	–1776	EK; H 1799–1803		Anno 1776 in EK	
25	27	G				HU ii, 60; WU 42		19
26	28	E♭				HU ii, 70; WU 43		19
27	29	F		1774	A (frag.)	HU ii, 82; WU 44		19
28	30	A				HU ii, 96; WU 45		19
29	31	E				HU ii, 106; WU 46		19, 52
30	32	b				HU ii, 116; WU 47		19, 22
31–6		6 Sonatas:	hpd/pf	OE, HV				
31	35	C		–31 Jan 1780		HU ii, 126; WU 48		
32	36	c♯		–31 Jan 1780 [c1770–75]		HU ii, 138; WU 49		46
33	37	D		–31 Jan 1780		HU ii, 146; WU 50		
34	38	E♭		–31 Jan 1780 [?c1770–75]		HU ii, 154; WU 51		
35	39	G		–8 Feb 1780		HU ii, 162; WU 52		46
36	20	c		177_	A (frag., incl. 3k)	HU ii, 174; WU 33		18, 71
37–9		[3] Sonatas:	hpd (/pf)					
37	43	A♭		–26 July 1783	—	HU iii, 1; WU 35	considered not authentic by Somfai (O1979, trans. 1995)	25
38	33	D		–17 Jan 1778	HV; H 1799	HU iii, 12; WU 34	date on MS in A-Wn	25
39	34	e		–15 Jan 1784		HU iii, 22; WU 53		25
40–42		3 Sonatas:	pf	–1784	HV; Haydn's ded. in 1st edn			
40	40	G				HU iii, 33; WU 54	also known in doubtful arr. for str trio	25
41	41	B♭				HU iii, 40; WU 55		25

No.	HXVI	Title, key	Instrument	Date	Authentication	Edition	Remarks	
42	42	D				HU iii, 48; WU 56		
43	48	Sonata, C	hpd (/pf)	−5 April [−?10 March] 1789	OE, HV	HU iii, 56; WU 58		25, 72
44	49	Sonata, E♭	pf	1789–[1 June] 1790	A	HU iii, 68; WU 59; facs. (Graz, 1982)		25, 29, 50, 53, 72
45	52	Sonata, E♭	pf	1794	A	HU iii, 84; WU 62		36, 72
46	50	Sonata, C	pf	[c1794/5]	H 1799–1803 (II only)	HU iii, 100; WU 60	earlier version of II appeared 1794 (WU iii, 121)	36, 72
47	51	Sonata, D	pf	[?c1794/5]	?OE	HU iii, 114; WU 61		36, 71

Note: sketch for inc. sonata, HU iii, 122; 'Sonata Pianoforte für den Nelson', mentioned in Elssler, *Haydn's vollendete Compositionen, A-Sm*, not verified, ? = H 19

Appendix W.1: Early harpsichord sonatas attributed to Haydn

No.	HXVI	Title, key	Date	Authentication	Edition	Remarks	
1	16	Divertimento, E♭	[?c1750–55]		HU i, 1	doubtful	
2	5	Divertimento, A	−1763 [?c1750–55]	H 1803	HU i, 6; WU 8	doubtful	
3	12	Divertimento, A	−1767 [?c1750–55]	H 1803	HU i, 14; WU 12	? I doubtful	
4	13	Parthia (Divertimento), E	−1767 [−?1760]	? H 1803	HU i, 19; WU 15	probably authentic; Haydn's statements in 1803 concerning his authorship were contradictory	71
5	2	Partita (Parthia), B♭	[−?1760]	—	HU i, 44; WU 11	probably authentic	
6	Es2 Add.	Parthia, E♭	[?c1755]	—	HU i, 53; WU 17	doubtful	
7	Es3 Add.	Parthia, E♭	[?c1764]	—	HU i, 60, 187; WU 18	doubtful; also attrib. Mariano Romano Kayser	
8	1	Partita (Divertimento), C	[?c1750–55]	H 1803	HU i, 68; WU 10	considered probably not authentic by Somfai (O1995)	
9	7	Partita (Parthia, Divertimento), C	−1766 [−?1760]	H 1803	HU i, 74; WU 2	probably authentic	
10	8	Parthia (Divertimento), G	−1766 [−?1760]	H 1803	HU i, 77; WU 3	probably authentic	
11	9	Divertimento, F	−1766 [−?1760]	H 1803	HU i, 80; WU 1	probably authentic; see Divertimento U 8	
12	10	Divertimento, C	−1767 [−?1760]	H 1803	HU i, 84; WU 6	probably authentic	
13a	G1	Divertimento, G	[−?1760]	—	HU i, 90; WU 4	probably authentic	
13b	11	Divertimento, G	−1767	H 1803	WU 5	?later combination of III of no.13a and 2 other movts; see appx X.I, 1–2	
14	XVII:D1	Variazione, D	?	—	HU i, 94; WU 7	3 movts: variations, minuet, finale; considered not authentic by Somfai (O1995)	

Note: see also appx U 8

Appendix W.2: Selected spurious sonatas

No.	H	Title, key	Date	Edition	Remarks
1	XVI:15	Sonata, C	–1785	M xiv/1, 80	arr. of Divertimento N 14; spurious though in Breitkopf & Härtel's Oeuvres de Haydn
2	XVI:17	Sonata, Bb	–1768	M xiv/1, 91	probably by J.G. Schwanenberger though authenticated by Haydn c1759–1803 according to Pohl
3	XVI:C1	Sonata, C	–1774	—	4 apparently heterogeneous movts; 1st movt by Liber
4	XVI:C2 Add.	Sonata, C	–1767	—	1 movt only; ? by Eckard
5	XVI:D1	Sonata militare (The conquest of Oczakow), D	17 Dec 1788–11 April 1789	Mw, xxxvi (1970), 53	by Kauer
6	i, 731	Sonata, Eb	–1789	(London, 1789)	with vn ad lib
7	XVI:B1	Sonata, Bb	?	—	
8	XVII:G2	Caprices (Fantasie et variations), G	–1787	(Paris, 1787)	? by Varhal
9–11		[3] Göttweiger Sonaten, C, A, D	?	(Wolfenbüttel, 1934)	by Hoffmeister
12	—	Children's Concerto (Concerto de bébé), C	?–1876	(London, 1964)	?19th-century forgery

X: MISCELLANEOUS KEYBOARD WORKS

Editions: *Joseph Haydn: Klavierstücke*, ed. S. Gerlach (Munich, 1969) [HU]
Joseph Haydn: Klavierstücke, ed. F. Eibner and G. Jarecki (Vienna, 1975) [WU] *Joseph Haydn: Tänze für Klavier*, ed. H.C.R. Landon and K.H. Füssl (Vienna, 1989) [WT]

No.	H	Title, key	Instrument	Date	Authentication	Edition	Remarks
1	XVII:1	Capriccio: Acht Sauschneider müssen sein, G	hpd	1765	A	HU, 5; WU, 1	72
2	XVII:2	20 Variazioni, A	hpd	–1771 [c1765]	EK	HU, 16; WU, 22 (in G)	first pubd 1788/9 as Arietta con 12 variazioni (WU, 41)
3	XVII:3	12 Variations, Eb	hpd	–1774 [c1770–74]	HV	HU, 28; WU, 33	theme arr. from minuet of str qt O 14
4	XVII:4	Fantasia (Capriccio), C	pf	[–?29 March] 1789	OE, HV	HU, 37; WU, 12	25, 7≡
5	XVII:5	6 Variations, C	pf	–9 Feb 1791 (?Nov 1790)	OE, HV	HU, 48; WU, 48	35, 7≡
6	XVII:6	Sonata (Un piccolo divertimento; Variations), f	pf	1793	A	HU, 54; WU, 33	
7	XXXIc:17b	(Untitled piece), D	(pf)	[?1791–5]	A (u)	HW v, 179; xx x/1, 97	written with song appx G.1, 2; ? country dance; cf L 21
8	XVIIa:1	Divertimento (Il maestro e lo scolare), F	hpd (4 hands)	–1778 [?c1768–70]	EK	WU, 78; (Mun.ch, 1982); (Vienna, 1996)	
9	XVII:9	Adagio, F	hpd/pf	–1786	Sk	HU, 68; WU, 69	sketch in PL-Kj

Note: for further works see pf trio V 20 (WU, 70) and pf sonata W 46

Appendix X.1: Selected works attributed to Haydn

No.	H	Title, key	Instrument	Date	Authentication	Edition	Remarks
1	XVI:11ii	Andante, g	hpd	−1767 [?c1755]	—	HW xviii/1, 181	extant as II of sonata appx W:1, 13b
2	XVI:IIiii	Minuet, G, Trio e	hpd	−1767 [?c1765]	—	HW xviii/1, 182	extant as III of appx W:1, 13b; trio doubtful; cf Q 26
3	XVII:D2 Add.	Allegro molto, D	hpd	[−?1765]	—	HW xviii/1, 184; D 533 (1976), acc. vn, b	frag., ?finale of sonata; in D 533 introduction to no.4; see appx V.2, 6
4	XVII:7	5 Variations, D	hpd	−1766 [?c1750–55]	H 1803	HU, 65; D 533 (1976), acc. vn, b	see above
5	IX:26	Minuetto, ♮♮	pf	−1785	—	HW xviii/1, 186	by Kirnberger, orig. in D, see HW v, preface
6	—						see X 9
7	—	Variationes, C	hpd/pf	?	—	—	variations on theme of aria, appx F.1, 4; orig. without author's name; see Schmid (O1970); cf also appx Y.4, 1; probably spurious
8	XVII:11*	Andante, C	hpd	−1807	—	(? Vienna, 1807)	not verified
9	XVII:12*	Andante con variazioni, Bb	pf	−1807	—	(Bryn Mawr, 1974)	probably spurious
10	XVIIa:2	Partita, F	hpd (4 hands)	[?c1768–70]	—	(Bryn Mawr, 1956)	doubtful
11–13	XVII:C2	3 Praeambula, C, C, G	org	?	—	(Hilversum, 1979)	
14	XVII:F2	Andante, F	org	?	—	(Hilversum, 1979)	probably spurious

Note: more works, probably spurious, listed in HXVII as C1 etc.

Appendix X.2: Arrangements

No.	H	Title, key	Date	Authentication	Edition	Remarks
1	IX:3	[12] Menuetti (with 4 Trios)	[c1765–7]	A (u)	HW v, 152; WT, 2	see L 2; cf appx X.3, 10
2	i, 799	Allegretto, G	1781–6	A (u)	HW xii/3, 189; WU, 76	arr. of finale of O 29
3	IX:8	XII Menuets (with 5 Trios)	−12 Jan 1785	—	HW v, 158; WT, 10	trio of minuet no.11 (HW v, 245) doubtful; cf L 10
4	IX:11	[12] Menuetti di ballo (with 11 Trios)	−22 Dec 1792	E	HW v, 165; WT, 21	arr. on request of the empress; cf L 18
5	IX:12	XII neue deutsche Tänze (with Trio and Coda)	−22 Dec 1792	A (u)	HW v 175; WT, 36	see L 19 and previous remark
6	iii, 302	(Largo assai), E	c1793	A (u frag.)	HW; xii/5, 223	arr. of II of O 59; only frags. extant
7	VIII:1–2	2 [Derbyshire] Marches, Eb, C (2 versions: 'A' and 'B')	c1795	A (u) (no.2, 'A'); RC (no.1), E (no.2, 'B') (both without author's name)	HW v, 226	see L 22

No.	H	Title, key	Date	Authentication	Edition	Remarks
8	i, 430	Variations sur le thème Gott erhalte den Kaiser, G	?1797–9	A (u)	WU, 64; (Munich, 1997)	arr. of II of O 62; later erroneously attrib. J. Gelinek

Note: arrs. of Loudon sym. (J 69) and the Seven Last Words (K 11) nor made but rev. Haydn; MS arr. of sym. J 96 described in *IMSCR VII: Cologne 1958*, 197, not autograph; authenticity of many printed arrs., incl. arr. of J 73, not verified; see also X 7

Appendix X.3: Selected doubtful and spurious arrangements

No.	H	Title, key	Date	Edition	Remarks
1	IX:9a	6 Minuetti	~Aug 1787	—	lost or unidentified
2	XVI:Es1	Sonata (Terzetto; Die Belagerung Belgrads), E♭	c1789–93	—	arr. (?by P. Pozelli) of terzeto Ξ 15c
3	IX:13	12 deutsche Tänze (with 5 Trios, Coda) aus dem k.k. Redouten Saale	1792 or later	—	first 2 pages of MS copy by J. Schellinger
4	IX:10	XII deutsche Tänze	~1793	(Mainz, 1937)	arrs. of melodies from opera L'arbore di Diana by Martín y Soler; probably = appx L 6
5	XVII:10	Allegretto, G	~1794	HW xxi, 49; WU, 74	arr. of piece for flute clock (Y 11)
6	IX:31 Add.	The Princess of Wales's Favorite Dance (Country Dance)	?1795	MT, cii (1961), 693	? part cf L 21
7	IX:28	[8] Zingarese	~21 April 1792	Strache xxvi (Vienna, 1930), 9	for date, see HW v, preface
8	IX:27	Ochsenmenuett (Menuet du bœuf)	c1805	(Mainz, n.d.)	taken from or gave rise to the following stage works: (?lost) vaudeville Le menuet du bœuf, ou Une leçon d'Haydn, 1805, by J.B. Constantin; (?lost) vaudeville Haydn, ou Le menuet de bœuf, 1812, by J.J. Gabriel and A.J.M. Wafflard; pasticcio Die Ochsenmenuett, 1823, see appx E 6
9a	i, 794	Variations, A	?	—	one of 5 different arrs. of variations in Q 2c
9b	XVII:8	Variations, D	?	—	see remark for no.9a
9c	i, 794	Variations, C	?	—	see remark for no.9a; differs widely from orig.
10	IX:20; XVII:F1	[18] Menuetti (with 7 Trios) and Aria	?	HW v, 193; aria also in HW xviii/1, 186	no.2 from appx X.2, 1, no.10; no.18 from sonata W 2; aria ? from lost early sonata
11	IX:21	[12] Menuets (with 5 Trios) de la redoute	?	(Zürich, n.d.)	
12	IX:22	Ballo redescho (10 deutsche Tänze)	?	Strache xxvi (Vienna, 1930), 5	listed in H as minuets
13	IX:29; IX:24	[5] Contredanze (Contredanse) (with Quadrille, Minuet)	?	Strache xxvi (Vienna, 1930), 5	
14	IX:30	Englischer Tanz	?	—	
15	i, 580	[3] Minuetti	?	Strache xxvii (Vienna, 1930), 5	from pf trios: appx V.1, 4, 8, 3
16	—	Fantaisie pour l'orchestre, d	?	(Paris, 1855)	arr. pf 4 hands by E.T. Eckhardt; see Mies (H1962)
17	IX:4a Add.	6 Minuetti (with 6 Trios)	?1770	—	arr. of minuets by M. Haydn, also attrib. Mozart κ 61f facs. in Landon (A1980), 638; composed Duncan McIntyre, see MT, cxxix (1988), 459
18a	—	Haydn's Strathspey, F	?c1795	in MT, cxxxvii (1986), 17	
18b	—	Haydn's Strathspey, E♭	?c1795	—	? composed George Jenkins, see HW v, preface

No.	H	Title, key	Date	Edition	Remarks
19	—	8 bars in 'The Pic Nic'	~1795	in MT, cxxix (1988), 460	arr. Nathaniel Gow
20	VIII:3/3bis	Marcia, E♭	?	HW v, 229	arr. of L 17
21	—	6 Sonatinas	?1961	(Budapest, 1961)	arrs. (by ? F. Brodszky) of baryton trios Q 56, 35, 72, 70, 34, 75

Y: WORKS FOR FLUTE CLOCK

Edition: *Joseph Haydn: Werke für das Laufwerk (Flötenuhr)*, ed. E.F. Schmid (Kassel, 1954) [S]

No.	hXIX	Title, key	Date	Authentication	No.: year of clock	Edition	Remarks
1	17	C	-1792 [?c1789]	A (u)	I: 1792	HW xxi, I.1; S 1	
2	10	Andante, C	-1792 [c1789]	A (u)	I: 1792; III: ?1796 (not 1772)	HW xxi, I.2; S 2	
3	18	Presto, C	-1792 [?c1789]	A (u)	I: 1792	HW xxi, I.3; S 3	2 versions, the longer one by P. Niemetz
4	16	Fuga, C	1789	A	II: 1793; III: ?1796 (not 1772)	HW xxi, II.7; S 24	
5	11	C	-1793 [?1789]	A	II: 1793; III: ?1796 (not 1772)	HW xxi, II.1; S 19	
6	12	Andante, C	-1793 [?1789]	A	II: 1793; III: ?1796 (not 1772)	HW xxi, II.2; S 20	
7	13	C	-1793 [?1789]	E (without author's name)	II: 1793; III: ?1796 (not 1772)	HW xxi, II.4; S 21	
8	14	C	-1793 [?1789]	E (without author's name)	II: 1793; III: ?1796 (not 1772)	HW xxi, II.5; S 22	
9	15	C	-1793 [?1789]	E (without author's name)	II: 1793; III: ?1796 (not 1772)	HW xxi, II.6; S 23	
10	31	Presto, C	[?1789]	A	—	HW xxi, II.3; S 31	
11	27	Allegretto, G	[?1793]	A (u)	II: 1793	HW xxi, III.1; S 27 cf appx X.3, 5	

Appendix Y.1: Adaptations for flute clock

No.	hXIX	Title, key	Date	Authentication	No.: year of clock	Edition	Remarks
1	9	Menuet, C	1788–92 [?c1789]	A (u)	I: 1792; III: ?1796 (not 1772)	HW xxi, I.4; S 11	uses III of O 42
2	28	Allegro, C	[1793]	A (u)	II: 1793	HW xxi, III.2; S 28	adapted from IV of O 55
3	29	C	[1793]	A (u)	II: 1793	HW xxi, III.3; S 29	adapted from III of J 101
4	30	Presto, G	1790–93 [?1793]	A (u)	II: 1793	HW xxi, II.4; S 30	adapted from IV of O 53
5	32	Allegro, F	1793 or later	A (u)	—	HW xxi, IV.1; S 32	adapted from IV of J 99

Appendix Y.2: Doubtful works

No.	HXIX	Title, key		No.: year of clock	Edition	Remarks
1	24	Presto, C		I: 1792	HW xxi, I.5; S 12	MS copy without author's name
2	21	C–F/G–C		I: 1792	HW xxi, appx A.4; S 7	no written source known
3	7	F/C		I: 1792; III: ?1796 (not 1772)	HW xxi, appx A.5; S 8	2 versions; no written source known
4	8	F/C		I: 1792; III: ?1796 (not 1772)	HW xxi, appx A.3; S 6	2 versions; no written source known
5	2	F		III: ?1796 (not 1772)	HW xxi, appx B.2; S 14	no written source known

Appendix Y.3: Doubtful adaptations

No.	HXIX	Title, key		No.: year of clock	Edition	Remarks
1	19	F/C		I: 1792	HW xxi, appx A.1; S 4	adapted from G 13; no written source known
2	20	F/C		I: 1792	HW xxi, appx A.2; S 5	uses III of J 85; no written source known
3	25	Marche, D		II: 1793	HW xxi, appx C.1; S 25	adapted from L 15, also combined with Beethoven woo 29 on clock by F.E. Azt
4	1	F		III: ?1796 (not 1772)	HW xxi, appx B.1; S 13	uses aria no.4 in E 17; no written source known
5	3	F		III: ?1796 (not 1772)	HW xxi, appx B.3; S 15	uses II of J 53; no written source known
6	5	F		III: ?1796 (not 1772)	HW xxi, appx B.5; S 17	adapted from II of Q 82; no written source known
7	6	F		III: ?1796 (not 1772)	HW xxi, appx B.6; S 18	adapted from II of Q 76; no written source known

Appendix Y.4: Spurious arrangements

No.	HXIX	Title, key		No.: year of clock	Edition	Remarks
1	22	F/C		I: 1792	HW xxi, appx A.6; S 9	aria by unidentified author (see appx F.1, 4); no written source known
2	23	F/C		I: 1792	HW xxi, appx A.7; S 10	from IV of H:C5, sym. by Dittersdorf; no written sources known
3	26	Andante, Allegro, E		II: 1793	HW xxi, appx C.2; S 26	Allegro: Aria alla polacca by ... Schuster; see E 26
4	4	C		III: ?1796 (not 1772)	HW xxi, appx B.4; S 16	Air or Dance russe by Giornovichi in Das Waldmädchen, ballet by P. Wranitzky and J. Kinsky (1796); orig. Russ an folksong 'Kamarinskaya', see Beethoven, woo71; no written source known

3. FOLKSONG ARRANGEMENTS

Z: ARRANGEMENTS OF BRITISH FOLKSONGS

Editions: *A Selection of Original Scots Songs*, compiled W. Napier, ii–iii (London, 1792–5) [N]

A Select Collection of Original Scotish Airs, compiled G. Thomson, i–iv (London and Edinburgh, 1802–5); v (London and Edinburgh, 1818); suppl. to v as *25 Additional Scottish Airs* (Edinburgh, 1826) [T]

The Select Melodies of Scotland, compiled G. Thomson, i, ii, v (London and Edinburgh, 1822); vi as *Thomson's Collection ... United to the Select Melodies of Scotland ... Ireland and Wales* (London and Edinburgh, 1824); suppl. as *20 Scottish Melodies* (Edinburgh, 1839) [TS]

A Select Collection of Original Welsh Airs, compiled G. Thomson, i–iii (London and Edinburgh, 1809–17) [TW]

A Select Collection of Original Irish Airs, compiled G. Thomson, i–iii (London and Edinburgh, 1814) [TI]

A Collection of Scotish Airs, compiled W. Whyte, i–ii (Edinburgh, 1804–7) [W]

for 1 voice unless otherwise stated; hXXXIa = Scottish, b = Welsh; key sometimes uncertain; dates based on research by I. Becker-Glauch

No.	HXXXI	Tune/Title, key	Accompaniment	Date	Authentication	Edition (no. = no. of piece except in HW)
1	a:131	Adieu to Llangollen, see Happiness lost	vn, bc	1795	HV	N iii, 31
		Ae fond kiss, e (?Celtic air)				
		Age & youth, see What can a young lassie do				
		Aileen a roon, see Robin Adair				
		Alas! Yat I came o'er the moor, see Last time I came o'er the muir				
2	b:48	Allurement of love, The, D	vn, vc, pf	1804	HV	—
		Anna, see Shepherds, I have lost my love				
		Answer, The, see My mither's ay glowran				
3	a:164	An thou wert mine ain thing, A	vn, vc, pf	1800	SC, HV	T iii, 20
4	a:164bis	An thou wert mine ain thing, A	vn, vc, pf	?1804	HV	W ii, 49
		An ye had been where I hae been, see Killicrankie				
		Argyll is my name, see Bannocks o' barley meal				
5	b:9	Ar hyd y nos, A (duet)	vn, vc, pf	1803	HV	TW i, 12
6	b:55	Aria di guerra e vittoria, D	vn, vc, pf	1804	HV	—
7	a:114	As I cam down by yon castle wa', e	vn, bc	1795	HV	N iii, 14
		As I came o'er the Cairney mount, see Old highland laddie				
		As Sylvia in a forest lay, see Maid's complaint				
8	a:184	Auld gudeman, The, B♭	vn, vc, pf (hpd)	1801	A (u), HV	T iii, 47
9	a:218	Auld lang syne, F	vn, vc, pf	1802/3	HV	W i, 24
10	a:168	Auld Robin Gray, D	vn, vc, hpd	1800	HV	T iii, 26
11	a:192	Auld Rob Morris, E♭ (duet)	vn, vc, hpd	1801	A (2nd version of coda only, u), HV	T i, 17
12	a:195	Auld wife ayont the fire, The, E♭	vn, vc, pf	1801	A (u)	T i, 39

No.	HXXXI	Tune/Title, key	Accompaniment	Date	Authentication	Edition (no. = no. of piece except in HW)
13	a:157	Ay waking, O!, E♭ (duet)	vn, vc, hpd	1800	HV	T iii, 11
14	a:57	Banks of Banna, The (?Irish air), see Shepherds, I have lost my love	vn, bc	~1792	HV	N ii, 57; HW xxxii/1, 60
15	a:171	Banks of Spey, The, C	vn, vc, pf	1801	HV	T iii, 29
		Banks of the Dee, The, see Langolée				
16	a:11	Bannocks o' barley meal, G (cf H 22)	vn, bc	~1792	HV	N ii, 11; HW xxxii/1, 11
17	a:11bis	Barbara Allen, d	vn, vc, hpd	1800	HV	T iii, 30
		Barbara Allen, c				
		Bashful lover [swain], The, see On a bank of flowers				
18	a:54	Be kind to the young thing, B♭	vn, bc	~1792	HV	N ii, 54; HW xxxii/1, 57
19	b:56	Bend of the horse shoe, The, B♭	vn, vc, pf	1804	HV	—
20	a:147	Bess and her spinning wheel, G	vn, bc	1795	HV	N iii, 47
21	a:178	Bessy Bell and Mary Gray, C	vn, vc, pf	1800	HV	N iii, 38
22	a:178bis	Bessy Bell and Mary Gray, C	vn, vc, pf	?1804	HV	W ii, 58
23	a:126	Bid me not forget, G	vn, bc	1795	HV	N iii, 26
24	a:58	Birks of Abergeldie, The, b	vn, bc	~1792	HV	N ii, 58; HW xxxii/1, 61
25	a:58bis	Birks of Abergeldie, The, b	vn, vc, hpd	1801	HV	T iii, 36
26	a:187	Birks of Invermay, The, G	vn, vc, pf	1801	HV	T i, 1
27	a:187bis	Birks of Invermay, The, G	vn, vc, pf	1802/3	—	W i, 1
		Black cock, The, see Ton y ceiliog du				
28	a:66	Black eagle, The, 𝄢	vn, bc	~1792	HV	N ii, 66; HW xxxii/1, 70
29	a:162	Blathrie o't, The, b	vn, vc, hpd	1800	HV	T iii, 19
30	a:68	Blink o'er the burn, sweet Betty, B♭	vn, bc	~1792	HV	N ii, 68; HW xxxii/1, 72
31	a:20	Blithsome bridal, The, D	vn, bc	~1792	HV	N ii, 20; HW xxxii/1, 20
32	a:20bis	Blithsome bridal, The, D	vn, vc, pf	1801	HV	T iv, 187
33	b:23	Blodau Llundain, C (duet)	vn, vc, pf	1804	HV	TW ii, 34
34	b:35	Blodau'r drain, g	vn, vc, pf	1805	HV	TW ii, 57
35	b:30	Blodau'r grug, C	vn, vc, pf	1805	HV	TW ii, 43
36	b:54	Blossom of the honey suckle, The, a	vn, vc, pf	1804	HV	—
		Blossom of the raspberry, The, see My jo Janet				
37	a:176	Blue bell[s] of Scotland, The, D (cf H 20)	vn, vc, pf/hpd	1801/2	HV	T iii, 35
38	a:39	Blue bonnets, C	vn, bc	~1792	HV	N ii, 39; HW xxxii/1, 42
39	a:246	Boatman, The, C	vn, vc, pf (hpd)	1801	A (u)	T iv, 183
40	a:101	Bonnie gray ey'd morn, The, B♭	vn, bc	1795	HV	N iii, 1
41	a:101bis	Bonnie gray ey'd morn, The, B♭	vn, vc, pf	1801	HV	T v, 224 (in G)
		Bonnie laddie, highland laddie, see Jingling Jonnie				
42	a:25	Bonniest lass in a' the warld, The, D	vn, bc	~1792	HV	N ii, 25; HW xxxii/1, 26
43	a:102	Bonnie wee thing, A	vn, bc	1795	HV	N iii, 2
44	a:102bis	Bonnie wee thing, A	vn, vc, pf	?1802/3	HV	W i, 28

No.	HXXXI	Tune/Title, key	Accompaniment	Date	Authentication	Edition (no. = no. of piece except in HW)
45	a:102ter	Bonnie wee thing, A (cf appx Z 1)	vn, vc, pf	1801	HV	TS i, 22 (no vn, vc)
		Bonnie Anne, see If a body meet a body				
		Bonny Barbara Allan, see Barbara Allen				
		Bonny black eagle, The, see Black eagle				
46	a:59	Bonny brucket lassie, The, D	vn, bc	–1792	HV	N ii, 59; HW xxxii/1, 62
47	a:172	Bonny Jean, D	vn, vc, pf	1800	HV	T iii, 31
		Bonny Jean, see Willie was a wanton wag				
48	a:94	Bonny Kate of Edinburgh, G	vn, bc	–1792	HV	N ii, 94; HW xxxii/1, 98
		Bonny, roaring Willie, see Rattling roaring Willy				
		Bonny Scot-man, The, see Boatman				
49	a:200	Braes of Ballenden, The, G	vn, vc, pf	1801	HV	T ii, 84
50	a:200bis	Braes of Ballenden, The, G	vn, vc, pf	1802/3	A (u)	W i, 27
51	a:207	Braes of Yarrow, The, A	vn, vc, pf	1802/3	A (u)	W i, 5
		Braw lads of Galla water, see Galla water				
		Bridegroom greets when the sun gangs down, The, see Auld Robin Gray				
		Bride's song, The, see Blithsome bridal				
52	a:46	Brisk young lad, The, g	vn, bc	–1792	HV	N ii, 46; HW xxxii/1, 49
53	a:46bis	Brisk young lad, The, e	vn, vc, hpd	1801	HV	T iv, 191
54	b:51	Britons, The, c	vn, vc, pf	1804	HV	—
55	a:170	Broom of Cowdenknows, The, Eb (with chorus 2vv)	vn, vc, hpd	1800	HV	T iii, 28
		Broom, the bonny broom, see Broom of Cowdenknows				
56	a:204	Bush aboon Traquair, The, Bb (duet)	vn, vc, pf	1802/3	—	W i, 2
		Busk ye, busk ye, see Braes of Yarrow				
		Butcher boy, The, see My Goddess woman				
		By the stream so cool and clear, see St Kilda song				
		Captain Cook's death, see Highland Mary				
57	a:224	Captain O'Kain, e (?Irish air)	vn, vc, pf	?1802/3	HV	W i, 37
		Captain's lady, The, see Mount your baggage				
		Carron side, see Frae the friends and land I love				
58	b:26	Castell Towryn, Eb	vn, vc, pf	1803	HV	TW ii, 38
		Cauld frosty morning, see Cold frosty morning				
59	a:55	Cauld kail in Aberdeen, D	vn, bc	–1792	HV	N ii, 55; HW xxxii/1, 58
60	a:55bis	Cauld kail in Aberdeen, D (duet)	vn, vc, pf	1801	A (u)	T i, 31
		Caun du delish, see Oran gaoil				
61	b:39	Cerdd yr hen-wr or coed, F (duet)	vn, vc, pf	1804	HV	TW iii, 67
		Charming highlandman, The, see Lewie Gordon				
62	b:12	Codiad yr haul, Bb (duet)	vn, vc, pf	1803	HV	TW i, 17
63	b:1	Codiad yr hedydd, Bb	vn, vc, pf	1803	HV	TW i, 1

No.	HXXXI	Tune/Title, key	Accompaniment	Date	Authentication	Edition (no. = no. of piece except in HW)
64	a:107	Cold frosty morning, A, F	vn, bc	1795	HV	N iii, 7
65	a:213	Collier's [bonnie] dochter, The, see Collier's bonny assie				
		Collier's bonny lassie, The, F	vn, vc, pf	?1802/3	HV	W i, 14
		Collier's lass, The, see Collier's bonny lassie				
66	a:97	Colonel Gardner, B♭	vn, bc	~1792	HV	N ii, 97; HW xxxii/1, 100
		Come kiss wi' me, come clap wi' me, see Now westlin winds				
		Comin thro' the rye, see If a body meet a body				
		Comin thro' the rye, see Auld lang syne				
67	b:21	Cornish May song, The, E♭	vn, vc, pf	1803	HV	TW ii, 31
68	a:216	Corn riggs, A (duet)	vn, vc, pf	1802/3	—	W i, 20
		Cornwallis's lament, see Sensibility				
69	a:144	Country lassie, A, D	vn, bc	1795	HV	N iii, 44
70	a:193	Craigieburn Wood, D	vn, vc, pf	1801	A (u)	T i, 32
		Crooked horn ewe, The, see Ewie wi' the crooked horn				
		Cuckoo, The [The cuckoo's nest], see I do confess thou art sae fair				
71	a:47	Cumbernauld House, E♭	vn, bc	~1792	HV	N ii, 47; HW xxxii/1, 50
72	b:4	Dafydd y Garreg-Wen, g	vn, vc, pf	1804	HV	TW i, 6
73	a:32	Dainty Davie, D	vn, bc	~1792	A (u, frag.); HV	N ii, 32; HW xxxii/1, 33
74	a:259	Day returns, The, E♭ (duet)	vn, vc, pf	?1804	HV	W ii, 47
75	a:136	Dear Silvia, E♭	vn, bc	1795	HV	N iii, 36
76	a:138	Death of the linner, The, E♭	vn, bc	1795	HV	N iii, 38
77	a:138bis	Death of the linner, The, D (duet)	vn, vc, hpd	1801	A (2nd version of coda only, u), HV	T iii, 39
		Deil's awa' wi' the exciseman, The, see Looking glass				
78	a:229	Deil tak the wars, B♭	vn, vc, pf	1801	HV	T iv, 157
79	b:59	Departure of the king, The, e	vn, vc, pf	1804	HV	—
80	b:34	Digan y pibydd coch, c (2 versions)	vn, vc, pf	1803	HV	TW ii, 56
81	a:217	Donald, B♭ (?Irish air)	vn, vc, pf	?1802/3	HV	W i, 21
82	a:139	Donald and Flora, D	vn, bc	1795	HV	N iii, 39
83	a:139bis	Donald and Flora, E♭	vn, vc, pf	1802/3	—	W i, 15
		Donocht Head, see Minstrel				
84	b:50	Door clapper, The, G (duet)	vn, vc, pf	1804	HV	—
85	b:14	Dowch i'r frwydr, B♭ (duet)	vn, vc, pf	1803	HV	TW i, 19
86	a:152	Down the burn, Davie, F (duet)	vn, vc, hpd	1800	HV	T iii, 3
		Drunken wife o' Galloway, The, see Hooly and fairly				
87	a:26	Duncan Davison, C	vn, bc	~1792	HV	N ii, 26; HW xxxii/1, 27

137

No.	HXXXI	Tune/Title, key	Accompaniment	Date	Authentication	Edition (no. = no. of piece except in HW)
88	a:34	Duncan Gray, G	vn, bc	-1792	HV	N ii, 34; HW xxxii/1, 35
		Duchess of Buccleugh's reel, see Sutor's daughter				
		Earl Douglas's lament, see Lady Randolph's complaint				
89	a:234	East Neuk o' Fife, The, F	vn, vc, pf	1801	HV	T iv, 165
		Eire a ruin, see Robin Adair				
90	a:74	Eppie Adair, e	vn, bc	-1792	HV	N ii, 74; HW xxxii/1, 78
91	b:27	Erddigan caer y waun, G	vn, vc, pf	1803	HV	TW ii, 39
92	a:203bis	Erin-go-bragh, C	vn, vc, pf	?1802/3	HV	W i, 16
93	b:20	Eryri wen, b	vn, vc, pf	1804	HV	TW i, 28
		Ettrick banks, see On Ettrick banks				
94	a:188	Ewe-bughts, The, d	vn, vc, pf	1801	HV	T i, 8
95	a:116	Ewie wi' the crooked horn, The, F	vn, bc	1795	HV	N iii, 16
96	a:116bis	Ewie wi' the crooked horn, The, G	vn, vc, hpd	1800	HV	T iii, 6
		Exile of Erin, The, see Erin-go-bragh				
		Failte na miosg, see My heart's in the highlands				
97	a:117	Fair Eliza, e (Gaelic air)	vn, bc	1795	HV	N iii, 17
98	a:236	Fair Helen of Kirkconnell, B♭	vn, vc, pf	1804	HV	T iv, 168
		Fairwell, thou fair day, see My lodging is on the cold ground				
99	a:156	Fee him, father, F	vn, vc, pf	1800	HV	T iii, 10
100	b:13	Ffarwel Ffranses, F	vn, vc, pf	1804	HV	TW i, 18
101	b:40	Ffarwel jeuengetid, E♭	vn, vc, pf	1804	HV	TW iii, 74
102	a:29	Fife and a' the lands about it, D	vn, bc	-1792	HV	N ii, 29; HW xxxii/1, 30
103	b:58	Flower of north Wales, The, C	vn, vc, pf	1804	HV	—
104	a:90	Flowers of Edinburgh, The, E♭	vn, bc	-1792	HV	N ii, 90; HW xxxii/1, 94
105	a:90bis	Flowers of Edinburgh, The, F	vn, vc, pf	1801	HV	—
106	a:212	Flowers of the forest, The, B♭	vn, vc, pf	?1802/3	HV	W i, 13
107	a:222	For the lack of gold, B♭	vn, vc, pf	?1802/3	HV	W i, 34
		14th of October, see Ye Gods!				
108	a:105	Frae the friends and land I love, E♭	vn, bc	1795	HV	N iii, 5
109	a:7	Fy! gar rub her o'er wi' strae, e	vn, bc	-1792	HV	N ii, 7; HW xxxii/1, 7
110	a:7bis	Fy! gar rub her o'er wi' strae, e (duet)	vn, vc, pf	1801	A (2nd version of introduction only, u), HV	T ii, 53
		Fy, let us a' to the bridal [wedding], see Blithsome bridal				
		Gaberlunzie (Gaberlunyie) man, The, see Brisk young lad				
111	a:179	Galashiels, E♭	vn, vc, hpd	1800	HV	T iii, 41
112	a:15	Galla water, D	vn, bc	-1792	HV	N ii, 15; HW xxxii/1, 15

No.	HXXXI	Tune/Title, key	Accompaniment	Date	Authentication	Edition (no. = no. of piece except in HW)
113	a:15bis	Galla water, D	vn, vc, pf	?1802/3	HV	W i, 30
114	a:15ter	Galla water, D	vn, vc, pf	1803	HV	—
115	a:45	Gardener's march, The, see Gard'ner wi' his paidle				
		Gard'ner wi' his paidle, The, A	vn, bc	–1792	HV	N ii, 45; HW xxxii/1, 48
		Gentle swain, The, see Johnny's gray breeks				
116	a:225	Gilderoy, g (duet)	vn, vc, pf	?1802/3	HV	W i, 39
117	a:196	Gil Morris [Morrice], Eb	vn, vc, hpd	1801	HV	T i, 45
		Gin you meet a bonny lassie, see Fy! gar rub her o'er wi' strae				
118	a:88	Glancing of her apron, The, D	vn, bc	–1792	HV	N ii, 88; HW xxxii/1, 52
		Gordons has [had] the guiding o't, The, see Strephon and Lydia				
119	b:2	Gorhoffedd gwyr Harlech, G	vn, vc, pf	1803	HV	TW i, 2
		Go to the ew-bughts, Marion, see Ewe-bughts				
120	a:13	Gramachree, Eb (Irish air)	vn, bc	–1792	HV	N ii, 13; HW xxxii/1, 13
121	a:13bis	Gramachree, D (Irish air)	vn, vc, pf	1801	HV	T i, 18
122	a:13ter	Gramachree, F (Irish air)	vn, vc, pf	?1802/3	HV	W i, 22
123	a:8	Green grow the rashes, d	vn, bc	–1792	HV	N ii, 8; HW xxxii/1, 8
124	a:8bis	Green grow the rashes, b (with chorus 2vv)	vn, vc, pf	1801	HV	T iv, 155
125	a:112	Green sleeves, e	vn, bc	1795	HV	N iii, 12
126	a:112bis	Green sleeves, e	vn, vc, pf	1801	A (frag.), HV	TS suppl., 150
127	b:15	Grisiel ground, Bb (duet)	vn, vc, pf	1803	HV	TW i, 21
		Had awa frae me, Donald, see Thou'rt gane awa'				
128	a:63	Hallow ev'n, D	vn, bc	–1792	HV	N ii, 63; HW xxxii/1, 66
129	b:42	Happiness lost, D	vn, vc, pf	1804	HV	TW iii, 86
		Happiness lost, see Tears that must e'er fall				
130	a:243	Happy topers, The, C (with chorus 2vv)	vn, vc, pf	1801	A (frag.), HV	T iv, 179
131	b:33	Hela'r ysgyfarnog, C	vn, vc, pf	1804	HV	TW ii, 50
		Hellwllyn, see Erin-go-bragh				
		Hemp-dresser, The, see Looking glass				
132	a:100	Her absence will not alter me, D	vn, bc	–1792	HV	N ii, 100; HW xxxii/1, 104
133	a:257	Here awa', there awa', F	vn, vc, pf	?1802/3	—	W ii, 45
134	a:49	Here's a health to my true love, b	vn, bc	–1792	HV	N ii, 49; HW xxxii/1, 52
		He's far away, see Weary pund o' tow				
		He who presum'd to guide the sun, see Maid's complaint				
		Hey now the day dawes, see Hey tutti taiti				
135	a:174	Hey tutti taiti, G	vn, vc, hpd	1801	HV	T iii, 33
		Highland lamentation, see Young Damon				
		Highland lassie [laddie], The, see Old highland laddie				

No.	hXXXI	Tune/Title, key	Acompaniment	Date	Authentication	Edition (no. = no. of piece except in HW)
136	a:159	Highland Mary, E♭	vn, vc, pf	1800	HV	T iii, 14
		Highway to Edinburgh, The, see Black eagle				
137	b:11	Hob y deri dando, G	vn, vc, pf	1803	HV	TW i, 16
138	b:16	Hob y deri danno, G	vn, vc, pf	1804	HV	TW i, 22
139	b:28	Hoffedd Hywel ab Owen Gwynedd, c	vn, vc, pf	1804	HV	TW ii, 40
		Hold away from me, Donald, see Thou'rt gane awa'				
140	a:237	Hooly and fairly, D	vn, vc, hpd	1801	HV	T iv, 170
		House of Glams, see Roslin Castle				
141	a:36	How can I be sad on my wedding day, D	vn, bc	-1792	HV	N ii, 36; HW xxxii/1, 38
142	a:67	How long and dreary is the night, D (Gaelic air)	vn, bc	-1792	HV	N ii, 67; HW xxxii/1, 71
		How sweet is the scene, see Humours o' glen				
		How sweet this lone vale, see Lone vale				
143	a:141	Hughie Graham, g	vn, bc	1795	HV	N iii, 41
144	a:256	Humours o' glen, The, a (?Irish air)	vn, vc, pf	?1802/3	HV	W ii, 44
145	a:140	I canna come ilke day to woo, A	vn, bc	1795	HV	N iii, 40
146	a:140bis	I canna come ilke day to woo, A	vn, vc, hpd	1801	HV	T v, 227
147	a:110	I do confess thou art sae fair, d	vn, bc	1795	HV	N iii, 10
148	a:87	I dream'd I lay, F	vn, bc	-1792	HV	N ii, 87; HW xxxii/1, 91
149	a:80	If a body meet a body, G	vn, bc	-1792	HV	N ii, 80; HW xxxii/1, 84
150	a:80bis	If a body meet a body, G	vn, vc, pf	1801	HV	T iii, 23
151	a:95	If e'er ye do well it's a wonder, D	vn, bc	-1792	HV	N ii, 95; HW xxxii/1, 99
152	a:17	I had a horse, b	vn, bc	-1792	HV	N ii, 17; HW xxxii/1, 17
153	a:17bis	I had a horse, c	vn, vc, pf	?1804	HV	W ii, 50
154	a:205	I'll never leave thee, D	vn, vc, pf	1802/3	—	W i, 3
155	a:3	I love my love in secret, G	vn, bc	-1792	HV	N ii, 3; HW xxxii/1, 3
156	a:30	I'm o'er young to marry yet, B♭	vn, bc	-1792	HV	N ii, 30; HW xxxii/1, 31
157	a:177	I wish my love were in a myre, B♭	vn, vc, hpd	1800	HV	T iii, 37
158	a:231	Jacobite air, A, B♭ (duet)	vn, vc, pf	1801	HV	T iv, 160
159	a:79	Jamie, come try me, D	vn, bc	-1792	HV	N ii, 79; HW xxxii/1, 83
160	a:132	Jenny drinks nae water, B♭	vn, bc	1795	HV	N iii, 32
161	a:252	Jenny's bawbee, G	vn, vc, pf (hpd)	1801	A (u)	T iv, 197
		Jenny's lamentation, see Jockie and Sandy				
162	a:99	Jenny was fair, E♭	vn, bc	-1792	HV	N ii, 99; HW xxxii/1, 102
163	a:263	Jingling Jonnie, F	vn, vc, pf	1801	HV	T ii (1817), 79 (in E♭)
		Jockey was the blythest lad, see Young Jockey				
164	a:91	Jockie and Sandy, G	vn, bc	-1792	HV	N ii, 91; HW xxxii/1, 95
		Jock the laird's brither, see Auld Rob Morris				
165	a:2	John Anderson, my jo, g	vn, bc	-1792	HV	N ii, 2; HW xxxii/1, 2
166	a:2bis	John Anderson, my jo, g	vn, vc, pf	?1802/3	HV	W i, 26
167	a:41	John, come kiss me now, E♭	vn, bc	-1792	HV	N ii, 41; HW xxxii/1, 44

No.	HXXXI	Tune/Title, key	Accompaniment:	Date	Authentication	Edition (no. = no. of piæx except in HW)
168	a:109	Johnie Armstrong, G	vn, bc	1795	HV	N iii, 9
169	a:154	Johnny's gray breeks, E♭/g	vn, vc, pf	1800	HV	T iii, 8
170	a:24	John of Badenyon, g	vn, bc	–1792	HV	N ii, 24; HW xxxii/1, 24
171	a:24bis	John of Badenyon, g	vn, vc, pf	1801	HV	T iv, 184
		Joyful widower, The, see Maggy Lauder				
172	a:220	Katherine Ogie, g	vn, vc, pf	1802/3	—	W i, 31
		Katy's answer, see My mither's ay glowran				
173	a:148	Kellyburn braes, E♭	vn, bc	1795	HV	N iii, 48
174	a:148bis	Kellyburn braes, D	vn, vc, hpd	1801	HV	T iv, 182
175	a:169	Killicrankie, C (cf H 25)	vn, vc, hpd	1801	HV	T iii, 27
		Kind Robin loves me, see Robin, quo' she				
		King James' march to Ireland, see Lochaber				
		Kirk wad let me be, The, see Blithsome brical				
		Lads of Leith, The, see She's fair and fause				
		Lady Badinscoth's reel, see My love she's but a lassie yet				
176	b:45	Lady Owen's delight [favourite], F	vn, vc, pf	1803	HV	—
177	a:127	Lady Randolph's complaint, G	vn, bc	1795	HV	N iii, 27
		Laird and Edinburgh Kate, The, see My mither's ay glowran				
178	b:43a	Lamentation of Britain, The, g (duet)	vn, vc, pf	1803	HV	—
179	b:43b	Lamentation of Cambria, The, g (duet)	vn, vc, pf	?1804	—	—
180	a:235	Langolee, G	vn, vc, pf	1801	HV	T iv, 167
181	b:57	La partenza dal paese e dalli amici, a	vn, vc, pf	1804	HV	—
		Lasses of the ferry, The, see Auld lang syne				
		Lass gin ye lo'e me, tell me now, see I canna come ilke day				
182	a:272	Lassie wi' the gowden hair, d	vn, vc, pf	1803	HV	—
183	a:23	Lass of Livingston, The, c	vn, bc	–1792	HV	N ii, 23; HW xxxii/1, 23
184	a:209	Lass of Lochroyan, The, a	vn, vc, pf	?1802/3	HV	W i, 7
185	a:160	Lass of Patie's mill, The, C	vn, vc, pf	1800	HV	T iii, 17
186	a:160bis	Lass of Patie's mill, The, C (duet)	vn, vc, pf	?1804	HV	W ii, 43
187	a:199	Last time I came o'er the muir, The, D	vn, vc, pf	1801	A (u)	T i, 80
188	a:199bis	Last time I came o'er the muir, The, D	vn, vc, pf	1802/3	A (u)	W i, 25
189	a:27	Leader haughs and yarrow, F	vn, bc	–1792	HV	N ii, 27; HW xxxii/1, 28
190	a:31	Lea-rig, The, F	vn, bc	–1792	HV	N ii, 31; HW xxxii/1, 32
191	a:31bis	Lea-rig, The, F	vn, vc, hpd	1800	E (without author's name)	T iv, 195 (in G)
192	a:31ter	Lea-rig, The, G	vn, vc, pf	?1802/3	HV	W i, 8
193	a:61	Let me in this ae night, d	vn, bc	–1792	HV	N ii, 61; HW xxxii/1, 64

No.	hXXXI	Tune/Title, key	Accompaniment	Date	Authentication	Edition (no. = no. of piece except in HW)
194	a:61bis	Let me in this ae night, d	vn, vc, pf	1801	HV	T iv, 156
195	a:215	Lewie Gordon, G	vn, vc, pf	?1802/3	HV	W i, 19
196	a:83	Lizae Baillie, F	vn, bc	–1792	HV	N ii, 83; HW xxxii/1, 87
197	b:7	Llwyn Onn, G	vn, vc, pf	1803	HV	TW i, 10
198	a:190bis	Lochaber, F	vn, vc, pf	?1804	HV	W ii, 60
199	a:163	Logan water, g	vn, vc, pf	1800	HV	T iii, 16
200	a:73	Logie of Buchan, g/B♭	vn, bc	–1792	HV	N ii, 73; HW xxxii/1, 77
201	a:175	Lone vale, The, B♭ (highland air)	vn, vc, hpd	1801	HV	T iii, 34
202	a:158	Looking glass, The, G	vn, vc, hpd	1801	HV	T iii, 13
		Loth to depart, see La partenza dal paese e dalli amici				
203	a:53	Love will find out the way, A	vn, bc	–1792	HV	N ii, 53; HW xxxii/1, 56
204	a:210	Low down in the broom, C	vn, vc, pf	1802/3	A (u)	W i, 11
		Lucky Nancy, see Dainty Davie				
205	a:182	Macpherson's farewell, A (with chorus, 2vv)	vn, vc, hpd	1801	HV	T iii, 44
		Madam Cossy, see Looking glass				
206	a:86	Maggie's rocher, e	vn, bc	–1792	HV	N ii, 86; HW xxxii/1, 90
207	a:55	Maggy Lauder, B♭	vn, bc	–1792	HV	N ii, 35; HW xxxii/1, 36
208	a:35bis	Maggy Lauder, A (cf H 24)	vn, vc, hpd	1800	HV	T iii, 25
209	a:35ter	Maggy Lauder, B♭	vn, vc, pf	?1804	HV	W ii, 64
		Maid in Bedlam, The, see Gramachree				
		Maid of Toro, The, see Captain O'Kain				
210	a:84	Maid's complaint, The, b	vn, bc	–1792	HV	N ii, 84; HW xxxii/1, 88
211	a:221	Maid that tends the goats, The, a	vn, vc, pf	?1802/3	HV	W i, 33
212	a:221bis	Maid that tends the goats, The, a	vn, vc, pf	1801	HV	T iv, 166
213	b:36	Maltraeth, G	vn, vc, pf	1804	HV	TW ii, 58
214	b:5	Mantell Siani, G	vn, vc, pf	1803	HV	TW i, 8
215	a:65	Marg'ret's ghost, D	vn, bc	–1792	HV	N ii, 65; HW xxxii/1, 68
		Margret's ghost, see William and Margaret				
216	b:49	Marsh of Rhuddlan, The, g	vn, vc, pf	1804	HV	—
217	a:1	Mary's dream, ♮	vn, bc	–1792	HV	N ii, 1; HW xxxii/1, 1
218	a:1bis	Mary's dream, ♮	vn, vc, hpd	1801	HV	T iii, 7
		McFarsence's testament, see Macpherson's farewell				
219	a:81	McGrigor of Rora's lament, C (Celtic air)	vn, bc	–1792	HV	N ii, 81; HW xxxii/1, 84
		McPherson's rant, see Macpherson's farewell				
220	b:6	Mentra Gwen, A (duet)	vn, vc, pf	1803	HV	TW i, 9
221	a:50	Merry may the maid be, d	vn, bc	–1792	HV	N ii, 50; HW xxxii/1, 53
222	a:50bis	Merry may the maid be, d (duet)	vn, vc, pf	?1804	HV	W ii, 56
		Miller, The, see Merry may the maid be				
		Miller's daughter, The, see If a body meet a body				
		Miller's wedding, The, see Auld lang syne				

No.	HXXXI	Tune/Title, key	Accompaniment	Date	Authentication	Edition (no. = no. of piece except in HW)
223	a:92	Mill, mill O!, The, Bb	vn, bc	-1792	HV	N ii, 92; HW xxxii/1, 96
224	a:92bis	Mill, mill O!, The, Bb	vn, vc, pf	?1802/3	HV	W ii, 42
225	a:115	Minstrel, The, c	vn, bc	1795	HV	N iii, 15
226	a:115bis	Minstrel, The, b	vn, vc, hpd	1801	HV	T iv, 186
		Miss Admiral Gordon's strathspey, see Poet's ain Jean				
		Miss Farquharson's reel, see My love she's but a lassie yet				
		Miss Hamilton's delight, see My jo Janet				
227	a:143	Morag, d (Celtic air)	vr, bc	1795	HV	N iii, 43
228	a:143bis	Morag, c (Celtic air)	vr, vc, pf	1801	HV	—
		Moudiewort, The, see O, for ane-and-twenty Tam!				
229	a:42	Mount your baggage, C	vr, bc	-1792	HV	N ii, 42; HW xxxii/1, 45
230	a:51	Mucking of Geordie's byer, The, e	vr, bc	-1792	HV	N ii, 51; HW xxxii/1, 54
231	a:51bis	Mucking of Geordie's byer, The, e	vr, vc, pf	1801	—	T ii, 66
232	a:242	Muirland Willy, d (with chorus 2vv)	vr, vc, pf	1801	A (2nd version of introduction only, u), HV	T iv, 177
		Musket salute The, see My heart's in the highlands				
233	b:31	Mwynen Cynwyd, Eb	vr, vc, pf	1804	HV	TW ii, 44
		My ain fireside, see Todlen hame				
		My ain kind deary, see Lea-rig				
234	a:189	My apron deary, G	vr, vc, pf	1801	A (u)	T i, 9
235	a:189bis	My apron deary, A	vr, vc, pf	?1802/3	HV	W i, 23
236	a:18	My boy Tammy, d	vr, bc	-1792	HV	N ii, 18; HW xxxii/1, 18
237	a:166	My dearie if thou die, e	vr, vc, pf	1800	HV	T iii, 22
238	a:120	My Goddess woman, c	vn, bc	1795	HV	N iii, 20
239	a:77	My heart's in the highlands, Bb (Celtic air)	vn, bc	-1792	HV	N ii, 77; HW xxxii/1, 80
		My Jockey was the blythest lad, see Young Jockey				
240	a:258	My jo Janet, C	vn, vc, pf	?1804	HV	W ii, 46
241	a:262	My lodging is on the cold ground, F (?Irish air)	vn, vc, pf	?1802/3	HV	W ii, 53
		My love's bonny when she smiles on me, see Flowers cf Edinburgh				
242	a:194	My love she's but a lassie yet, C (cf H 21)	vn, vc, pf	1801	A (u)	T i, 35
		My love's in Germanie, see Wish				
		My Mary, dear departed shade, see Highland Mary				
243	a:70	My mither's ay glowran o'er me, e	vn, bc	-1792	HV	N ii, 70; HW xxxii/1, 74
244	a:70bis	My mither's ay glowran o'er me, e	vn, vc, pf	1800	HV	T iv, 194b
245	a:37	My Nanie, O, c	vn, bc	-1792	HV	N ii, 37; HW xxxii/1, 39
246	a:37bis	My Nanie, O, c	vn, vc, pf	1802/3	—	W i, 17
247	a:37ter	My Nanie, O, c	vn, vc, pf	1803	HV	—

No.	HXXXI	Tune/Title, key	Acompaniment	Date	Authentication	Edition (no. = no. of piece except in HW)
248	a:37quater	My Nanie, O, c (duet)	vn, vc, pf	1801	HV	T i (1822), 4
		My plaid away, see O'er the hills and far away				
		Nancy's to the green-wood gane, see Scornfu' Nancy				
		Nanny, O, see My Nanie, O				
		Nelly's dream, see Marg'ret's ghost				
		New hilland laddie, see Lass of Livingston				
249	b:60	New year's gift, The, a	vn, vc, pf	1804	HV	—
		Nine pint cogie, see Collier's bonny lassie				
250	a:125	Nithsdall's welcome hame, D	vn, bc	1795	HV	N iii, 25
251	b:29	Nos galan, G	vn, vc, pf	1803	HV	TW ii, 41
252	a:111	Now westlin winds, c	vn, bc	1795	HV	N iii, 11
253	a:89	O bonny lass, e (?Irish air)	vn, bc	-1792	HV	N ii, 89; HW xxxii/1, 93
254	a:48	O can ye labor lea, see Auld lang syne				
		O can you sew cushions, G	vn, bc	-1792	HV	N ii, 48; HW xxxii/1, 51
255	a:16	O'er bogie, g	vn, bc	-1792	HV	N ii, 16; HW xxxii/1, 16
256	a:16bis	O'er bogie, ♮	vn, vc, pf	1801	HV	T iii, 40
257	a:149	O'er the hills and far away, ♯ or b	vn, bc	1795	HV	N iii, 49
258	a:149bis	O'er the hills and far away, B♭ (with chorus 2vv)	vn, vc, pf	1801	HV	T iv, 161
259	a:122	O'er the moor amang the heather, E♭	vn, bc	1795	HV	N iii, 22
260	a:122ter	O'er the moor amang the heather, E♭	vn, vc, pf	?1802/3	HV	W i, 32
		Of a' the airts, see Poet's ain Jean				
		Of noble race was Shenkin, see Y gadly's				
261	a:108	O, for ane-and-twenty Tam!, E♭	vn, bc	1795	HV	N iii, 8
262	a:85	Oh, onochrie [Oh! ono Chrio], F (Irish air)	vn, bc	-1792	HV	N ii, 85; HW xxxii/1, 89
		Oh, open the door, Lord Gregory, see Lass of Lochroyan				
263	a:248	Old highland laddie [lassie], The, D	vn, vc, pf	1801	HV	T iv, 189
		Old man, The, see My jo Janet				
		O let me in this ae night, see Let me in this ae night				
264	a:142	On a bank of flowers, c	vn, bc	1795	HV	N iii, 42
265	a:151	On Ettrick banks, D	vn, vc, hpd	1800	HV	T iii, 1
		On the death of Delia's linnet, see Death of the linnet				
266	a:249	Oonagh [Oonagh's waterfall], d (Irish air)	vn, vc, pf	1801	HV	T iv, 190
267	a:255	Open the door, E♭ (?Irish air)	vn, vc, pf	?1804	HV	W ii, 41
268	a:228	O poortith cauld, see I had a horse				
		Oran gaoil, d (duet; Gallic air)	vn, vc, pf	1801	HV	T iv, 154
		O saw ye my father, see Saw ye my father				
		O steer her up and had her gaun, see Steer her up				
		Palmer, The, see Open the door				
269	b:22	Pant corlant yr wyn: neu, Dafydd or Garreg-las, B♭ (duet)	vn, vc, pf	1804	HV	TW ii, 33

No.	HXXXI	Tune/Title, key	Accompanimen	Date	Authentication	Edition (no. = no. of piece except in HW)
270	a:241	Pat & Kate, B♭ (duet; Irish air)	vn, vc, pf	1803	HV	T iv, 175
271	a:167	Peggy, I must love thee, G (duet)	vn, vc, hpd	1801	A (2nd version of coda only, u)	T iii, 24
272	a:96	Peggy in devotion, C	vn, bc	–1792	HV	N ii, 96; HW xxxii/1, 100
273	a:33	Pentland Hills, F	vn, bc	–1792	HV	N ii, 33; HW xxxii/1, 34
		Phely & Willy, see Jacobite air				
		Phoebe, see Yon wild mossy mountains				
274	a:183	Pinkie House, D	vn, vc, pf	1800	HV	T iii, 46
275	a:10	Ploughman, The, D	vn, bc	–1792	HV	N ii, 10; HW xxxii/1, 10
276	a:230	Poet's ain Jean, The, G	vn, vc, pf	1801	HV	T iv, 159
277	a:230bis	Poet's ain Jean, The, A (duet)	vn, vc, pf	?1804	HV	W ii, 66
278	a:265	Polwarth on the green, B♭ (duet)	vn, vc, hpd	1801	HV	T v, 218
279	b:53	Poor pedlar, The, B♭	vn, vc, pf	1804	HV	—
280	a:113	Posie, The, c	vn, bc	1795	HV	N iii, 13
281	b:52	Pursuit of love, The, D	vn, vc, pf	1804	HV	—
282	a:161	Queen Mary's lamentation, The, E♭	vn, vc, pf	1800	HV	T iii, 18
		Ranting highlandman, The, see White cockade				
		Ranting, roving Willie, see Rattling roaring Willy				
283	a:227	Rattling roaring Willy, F	vn, vc, pf (hpd)	1801	A (u)	T iv, 153
		Raving winds, see McGrigor of Rora's lament				
284	b:38	Reged, G	vn, vc, pf	1804	HV	TW ii, 60
285	b:8	Rhyfelgyrch Cadpen Morgan, B♭	vn, vc, pf	1803	HV	TW i, 11
286	a:202	Robin Adair, C (duet; ?Irish air)	vn, vc, pf	1801	A (u)	T ii, 52
		Robin is my only jo, see Robin, quo' she				
287	a:72	Robin, quo' she, G	vn, bc	–1792	HV	N ii, 72; HW xxxii/1, 76
288	a:72bis	Robin, quo' she, G	vn, vc, pf	?1804	HV	W ii, 48
		Roger's farewell, see Auld lang syne				
		Rory Dall's port, see Ae fond kiss				
289	a:135	Rose bud, The, B♭	vn, bc	1795	HV	N iii, 35
290	a:191	Roslin Castle [Roslane Castle], c	vn, vc, pf	1801	HV	T i, 14
291	a:191bis	Roslin Castle [Roslane Castle], c	vn, vc, pf	1802/3	A (u)	W i, 9
292	a:165	Rothiemurche's rant, C	vn, vc, pf	1800	A (u)	T iii, 21
		Row saftiy, thou stream, see Captain O'Kain				
293	a:103	Roy's wife of Aldivalloch, C	vn, bc	1795	HV	N iii, 3
294	a:223	Sae merry as we ha'e been, C	vn, vc, pf	1802/3	A (u)	W i, 35
		Sandie and Jockie, see Jockie and Sandy				
		Sawney will never be my love again, see Corn riggs				
		Sawnie's pipe, see Colorel Gardner				
		Saw ye Johnnie cummin? quo' she, see Fee him, father				

No.	HXXXI	Tune/Title, key	Acompaniment	Date	Authentication	Edition (no. = no. of piece except in HW)
295	a:5	Saw ye my father?, D	vn, bc	−1792	HV	N ii, 5; HW xxxii/1, 5
296	a:5bis	Saw ye my father?, D (cf H 23)	vn, vc, hpd	1800	HV	T iii, 2
297	a:5ter	Saw ye my father?, D	vn, vc, pf	?1804	HV	W ii, 51
298	a:56	Saw ye nae my Peggy?, d	vn, bc	−1792	HV	N ii, 56; HW xxxii/1, 59
299	a:185	Scornfu' Nancy, B♭	vn, vc, pf	1800	HV	T iii, 48
300	a:185bis	Scornfu' Nancy, B♭	vn, vc, pf	?1804	HV	W ii, 55
301	a:173	Scots Jenny, see Jenny was fair Sensibility, E♭	vn, vc, pf	1800	HV	T iii, 32
		Seventh of November, see Day returns				
		She grip'd at the greatest on't, see East Neuk o' Fife				
302	a:21	Shepherd Adonis, The, g	vn, bc	−1792	HV	N ii, 21; HW xxxii/1, 21
303	a:93	Shepherds, I have lost my love, D	vn, bc	−1792	HV	N ii, 93; HW xxxii/1, 97
304	a:93bis	Shepherds, I have lost my love, C	vn, vc, pf	?1802/3	HV	W ii, 54
305	a:106	Shepherd's son, The, G	vn, bc	1795	HV	N iii, 6
306	a:106bis	Shepherd's son, The, G	vn, vc, pf	?1802/3	HV	W i, 40
307	a:106ter	Shepherd's son, The, A (2 versions)	vn, vc, pf	1804	HV	TS ii, 4 (no vn, vc)
308	a:128	Shepherd's wife, The, E♭	vn, bc	1795	HV	N iii, 28
309	a:128bis	Shepherd's wife, The, E♭	vn, vc, pf	1801	HV	T iii, 12
310	a:219	She rose and loot me in, d	vn, vc, pf	?1802/3	HV	W i, 29
311	a:219bis	She rose and loot me in, d	vn, vc, pf	1801	HV	—
		She says she lo'es me best of a', see Oonagh				
312	a:121	She's fair and fause, e	vn, bc	1795	HV	N iii, 21
313	a:208	Silken snood, The, E♭	vn, vc, pf	1802/3	—	W i, 6
314	a:260	Siller crown, The, F	vn, vc, pf	?1804	HV	W ii, 53
		Sir Alex. Don, see Auld lang syne				
315	a:250	Sir Patrick Spence, A	vn, vc, pf	1803	HV	T iv, 193
316	a:137	Slave's lament, The, d	vn, bc	1795	HV	N iii, 37
317	a:44	Sleepy bodie, F	vn, bc	−1792	HV	N ii, 44; HW xxxii/1, 47
		So for seven years, see Tho' for sev'n years				
318	a:60	Soger laddie, The, E♭	vn, bc	−1792	HV	N ii, 60; HW xxxii/1, 63
319	a:60bis	Soger laddie, The, E♭	vn, vc, pf	1801	HV	T iv, 172
		Soldier laddie, The, see Soger laddie				
		Soldier's dream, The, see Captain O'Kain				
		Soldier's return, The, see Mill, mill O!				
320	a:78	Steer her up, and had ger gawin, B♭	vn, bc	−1792	HV	N ii, 78; HW xxxii/1, 82
321	a:19	St Kilda song, F	vn, bc	−1792	HV	N ii, 19; HW xxxii/1, 19
322	a:145	Strathallan's lament, D	vn, bc	1795	HV	N iii, 45
323	a:145bis	Strathallan's lament, D	vn, vc, pf	1801	HV	T iv, 178
324	a:150	Strephon and Lydia, E♭	vn, bc	1795	HV	N iii, 50
		Sun had loos'd his weary team, The, see Looking glass				

No.	HXXXI	Tune/Title, key	Accompaniment	Date	Authentication	Edition (no. = no. of piece except in HW)
325	a:198	Sutor's daughter, The, G (duet)	vn, vc, pf	1801	A (u)	T ii, 77
326	a:261	Sweet Annie, g	vn, vc, pf	?1802/3	HV	W ii, 62
327	b:44	Sweet melody of north Wales, The, Bb	vn, vc, pf	1803	HV	—
		Sweet's the lass that loves me, see Bess and her spinning wheel				
328	a:180	Tak' your auld cloak about ye, g	vn, vc, pf	1800	HV	T ii, 42
329	a:180bis	Tak' your auld cloak about ye, g	vn, vc, pf	?1804	HV	W ii, 57
		Tarry woo', see Lewie Gordon				
330	a:123	Tears I shed, The, e	vn, bc	1755	HV	N iii, 23
331	a:201	Tears of Caledonia, The, d	vn, vc, pf	1801	HV	T ii, 87
332	a:186	Tears that must ever fall, D	vn, vc, pf	1801	HV	T iii, 49
		Their groves o' sweet myrtle, see Humours o' gler				
333	a:14	This is no mine ain house, Bb	vn, bc	–1792	HV	N ii, 14; HW xxxii/1, 14
334	a:14bis	This is no mine ain house, Bb	vn, vc, pf	?1802/3	HV	W i, 38
335	a:146	Tho' for sev'n years and mair, F	vn, bc	1795	HV	N iii, 46
336	a:12	Thou'rt gane awa', A	vn, bc	–1792	HV	N ii, 12; HW xxxii/1, 12
337	a:12bis	Thou'rt gane awa', A	vn, vc, pf	?1802/3	HV	W i, 36
338	a:264	Three captains, The, Eb (Irish air)	vn, vc, pf	1803	HV	T iv (1817), 193
339	a:181	Thro' the wood, laddie, F	vn, vc, hpd	1800	HV	T iii, 43
340	a:52	Tibby Fowler, b	vn, bc	–1792	HV	N ii, 52; HW xxxii/1, 55
		'Tis woman, see Bonnie gray ey'd morn				
341	a:130	Tither morn, The, F	vn, bc	1795	HV	N iii, 30
342	a:98	To daunton me, d	vn, bc	–1792	HV	N ii, 98; HW xxxii/1, 102
343	a:6	Todlen hame, A	vn, bc	1792	HV	N ii, 6; HW xxxii/1, 6
344	a:6bis	Todlen hame, A	vn, vc, pf	?1802/3	HV	W i, 61
345	b:18	Ton y ceiliog du, Bb (duet)	vn, vc, pf	1804	HV	TW i, 24
346	b:3	Torriad y dydd, b	vn, vc, pf	1803	HV	TW i, 4
		To the rose bud, see Rose bud				
		Tranent Muir, see Killicrankie				
347	b:41	Troiad y droell, Bb (duet)	vn, vc, pf	1804	HV	TW iii, 75
348	b:17	Tros y garreg, g	vn, vc, pf	1804	HV	TW i, 23
349	a:206	Tweedside, G (duet)	vn, vc, pf	1802/3	A (u)	W i, 4
350	b:10	Twll yn ei boch, C	vn, vc, pf	1803	HV	TW i, 14
351	a:233	Up and war them a' Willy, F	vn, vc, pf	1801	HV	T iv, 163
352	a:28	Up in the morning early, g	vn, bc	–1792	HV	N ii, 28; HW xxxii/1, 29
353	a:28bis	Up in the morning early, g	vn, vc, pf	?1802/3	HV	W ii, 52
354	a:28ter	Up in the morning early, g	vn, vc, pf	1801	HV	—
355	a:133	Vain pursuit, The, C	vn, bc	1795	HV	N iii, 33
356	a:9	Waefu' heart, The, F	vn, bc	–1792	HV	N ii, 9; HW xxxii/1, 9

No.	HXXXI	Tune/Title, key	Accompaniment	Date	Authentication	Edition (no. = no. of piece except in HW)
357	a:9bis	Waefu' heart, The, F	vn, vc, pf	?1802/3	A (Sk), HV	W i, 10
358	a:214	Waly, waly, D	vn, vc, pf	1802/3	A (u)	W i, 18
359	a:214bis	Waly, waly, D (cf appx Z 30)	vn, vc, pf	1801	HV	—
		Wandering Willie, see Here awa'				
		Wap at the widow, my laddie, see Widow				
360	a:69	Wat ye wha I met yestreen?, E♭	vn, bc	–1792	HV	N ii, 69; HW xxxii/1, 73
361	a:69bis	Wat ye wha I met yestreen?, e	vn, vc, pf	1801	HV	T iv, 194a
362	a:40	Wauking of the fauld, The, D	vn, bc	–1792	HV	N ii, 40; HW xxxii/1, 42
363	a:129	Weary pund o' tow, The, G	vn, bc	1795	HV	T iii, 4
364	a:129bis	Weary pund o' tow, The, F	vn, vc, hpd	1801	HV	N iii, 29
365	a:124	Wee wee man, The, F	vn, bc	1795	HV	N iii, 24
366	a:124bis	Wee wee man, The, E♭	vn, vc, pf	1801	HV	T iii, 15
		Welcome home, old Rowley, see Thou 'rt gane awa'				
367	a:244	What ails this heart of mine, g (duet)	vn, vc, pf	1804	SC, HV	T iv, 180
368	a:134	What can a young lassie do, b	vn, vc, pf	1795	HV	N iii, 34
369	a:134bis	What can a young lassie do, b (with chorus 2vv)	vn, vc, pf	1801	HV	T iii, 45
		What shall I do with an auld man, see What can a young lassie do				
370	a:62	When she came ben she bobbit, e	vn, bc	–1792	HV	N ii, 62; HW xxxii/1, 65
		Where Helen lies, see Fair Helen of Kirkconnell				
371	a:104	While hopeless, e	vn, bc	1795	HV	N iii, 4
372	a:76	Whistle o'er the lave o't, F	vn, bc	–1792	HV	N ii, 76; HW xxxii/1, 80
373	a:76bis	Whistle o'er the lave o't, F	vn, vc, pf (hpd)	1801	A (u)	T iv, 169
374	a:22	White cockade, The, D	vn, bc	–1792	HV	N ii, 22; HW xxxii/1, 22
375	a:118	Widow, The, E♭	vn, bc	1795	HV	N iii, 18
376	a:75	Widow, are ye waking?, E♭	vn, bc	–1792	HV	N ii, 75; HW xxxii/1, 79
377	a:75bis	Widow, are ye waking?, E♭	vn, vc, pf	?1804	HV	W ii, 59
378	a:153	William and Margaret, g	vn, vc, hpd	1800	HV	T iii, 5
		Willie brew'd a peck o' maut, see Happy topers				
379	a:4	Willie was a wanton wag, C	vn, bc	–1792	HV	N ii, 4; HW xxxii/1, 4
380	a:4bis	Willie was a wanton wag, B♭	vn, vc, pf	1801	HV	T iv, 152
381	b:47	Willow hymn, The, d	vn, vc, pf	1803	HV	—
		Will ye go to Flanders, see Gramachree				
382	a:82	Willy's rare, B♭	vn, bc	–1792	HV	N ii, 82; HW xxxii/1, 86
		Wilt thou be my dearie, see Sutor's daughter				
383	b:46	Winifreda, E♭ (duet)	vn, vc, pf	1803	HV	—
384	a:245	Wish, The, g	vn, vc, pf	1801	HV	T iv, 181
		Wo betyd thy wearie bodie, see Bonnie wee thing				
		Woes my heart that we shou'd sunder, A (duet), see Black eagle				
385	a:155	Women's work will never be done	vn, vc, hpd	1800	HV	T iii, 9

No.	HXXXI	Tune/Title, key	Accompaniment	Date	Authentication	Edition (no. = no. of piece except in HW)
386	a:38	Woo'd and married and a', d	vn, bc	–1792	HV	N ii, 38; HW xxxii/1, 40
387	a:38bis	Woo'd and married and a', d (with chorus, 2vv)	vn, vc, pf	1801	A (2nd version of coda only, u),	T iii, 50
388	b:19	Wyres Ned Puw, g	vn, vc, pf	1804	HV	TW i, 26
389	b:25	Y bardd yn ei awen, C	vn, vc, pf	1804	HV	TW ii, 36
390	b:32	Y Cymry dedwydd, Bb	vn, vc, pf	1804	HV	TW ii, 48
391	a:43	Ye Gods! was Strephon's picture blest, D	vn, bc	–1792	HV	N ii, 43; HW xxxii/1, 46
392	a:211	Yellow hair'd laddie, The, D (duet)	vn, vc, pf	1802/3	—	W i, 12
393	b:24	Y gadly's c (duet)	vn, vc, pf	1803	HV	TW ii, 35
394	a:119	Yon wild mossy mountains, g	vn, bc	1795	HV	N ii, 19
395	a:71	Young Demon, Bb	vn, bc	–1792	HV	N ii, 71; HW xxxii/1, 75
		Young highland rover, The, see Morag				
396	a:64	Young Jockey was the blythest lad, a	vn, bc	–1792	HV	N ii, 64; HW xxxii/1, 67
397	a:64bis	Young Jockey was the blythest lad, a	vn, vc, pf	1801	HV	TS suppl, 50
		Young laird and Edinburgh Katy, The, see Wat ye wha I met yestreen?				
		Young Peggy blooms, see Boatman				
398	b:37	Yr hen erddigan, c	vn, vc, pf	1803	HV	TW ii, 59

Note: 4 Scotch Songs, written in London, 1791–5 (Gr, Dies), lost or unidentified

Some settings of 1803 and later, doubtful: F. Kalkbrenner, during his stay with Haydn, 'was employed upon many of those popular Scottish airs, which are published by Mr. Thompson, of Edinburgh', see 'Memoir of Mr. Frederick Kalkbrenner' in Walter (E(i)1982)

Appendix Z: Doubtful and spurious settings

No.	HXXXI	Tune/Title, key	Accompaniment	Date	Edition	Remarks
1	a:102quater	Bonnie wee thing, A	pf	?	TS vi, 22	arr. of Z 45 for 3vv by Beethoven, woo153c, no.4
2	a:232	Border widow's lament, The, A	vn, vc, pf	1805	T iv, 162	by Neukomm
3	a:226	Braes of Ballochmyle, The, Eb	vn, vc, pf	1805	T iv, 151	by Neukomm
4	a:224bis	Captain O'Kain, e (?Irish air)	vn, vc, pf	1805	—	by Neukomm
		Colin to Flora, see Rock and a wee pickle tow				
		Come under my plaidy, see Johny MacGill				
5	a:253A	Cro Challin, F	vn, vc, pf	1805	T iv, 198	by Neukomm
6	a:203	Erin-go-bragh, C	vn, vc, pf	1805	T ii, 98	by Neukomm
		Exile of Erin, The, see Erin-go-bragh				
		Get up and bar the door, see Rise up and bar the door				

No.	HXXXI	Tune/Title, key	Acompaniment	Date	Edition	Remarks
		Good night, and God be with you, see Good night and joy be wi' ye a'				
7	a:254	Good night and joy be wi' ye a', G	vn, vc, pf	1803	T iv, 200	by Neukomm
8	a:63bis	Hallow ev'n, D	vn, vc, pf	1803	T v, 225	by Neukomm
9	a:247	Happy Dick Dawson, D	vn, vc, pf	1803	T iv, 185	by Neukomm
10	a:257bis	Here awa', there awa', d (duet)	vn, vc, pf	1803	—	?by Neukomm, see Angermüller (E(i)1974)
		I loe na a laddie but ane, see Happy Dick Dawson				
		Jenny beguil'd the webster, see Jenny dang the weaver				
11	a:240	Jenny dang the weaver, B♭	vn, vc, pf	1803	T iv, 174	by Neukomm
12	a:251	Johny Faw, B♭	vn, vc, pf	1804	T iv, 196	?by Neukomm, see Haydn's letter of 3 April 1804; altered version signed by Haydn
13	a:238	Johny MacGill, E♭ (?Irish air)	vn, vc, pf	1803	T iv, 171	by Neukomm
14	a:269	Kelvin Grove, G	pf	?	TS vi, 30	doubtful
15	a:190	Lochaber, F	vn, vc, pf	1803	T i, 10	?by Neukomm, see Angermüller (E(i)1974)
16	a:81bis	McGrigor of Rora's lament, C	vn, vc, pf	1803	T iv, 176	by Neukomm
17	a:268	My love's a wanton wee thing, D	vn, vc, pf	1803	TS vi, p.44 (no vn, vc)	by Neukomm
		My silly auld man, see Johny MacGill				
		My wife's a wanton, wee thing, see My love's a wanton wee thing				
18	a:89bis	O bonny lass, E♭	vn, vc, pf	1803	T iv, 164	by Neukomm
19	a:122bis	O'er the moor amang the heather, E♭	vn, vc, pf	1803	T iv, 158	by Neukomm
20	a:273	O gin my love were yon red rose, a	vn, vc, pf	1804	—	?by Neukomm, see Haydn's letter of 3 April 1804
21	a:267	Over the water to Charlie, D	vn, vc, pf	1803	TS vi, p.36 (no vn, vc)	by Neukomm
22	a:271	O were my love yon lilac fair, a	pf	?	TS vi, 32	doubtful
23	b:61	Parson boasts of mild ale, The, g (Irish air)	vn, vc, pf	1803	TI i, 30	by Neukomm
24	a:197	Rise up and bar the door, F	vn, vc, pf	1803	T i, 47	by Neukomm
25	a:253B	Rock and a wee pickle tow, The, F	vn, vc, pf	1803	T iv, 199	by Neukomm
26	a:266	Sailor's lady, The, A	pf	?	TS v, 37	doubtful
		Savourna deligh (Irish air), see Erin-go-bragh				
27	a:239	Shelah O'Neal, F	vn, vc, pf	1803	T iv, 173	by Neukomm
		Tibbie Dunbar, see Johny MacGill				
28	a:52bis	Tibby Fowler, b	vn, vc, pf	1803	T iv, 192	by Neukomm
29	a:270	Tullochgorum, D	vn, vc, pf	1803	T v (suppl.), 246	by Neukomm
		Waes me for Prince Charlie, see Johny Faw				
30	a:214ter	Waly, waly, D	vn, vc, pf	?	T i (1822), 19	doubtful duet arr. of Z 359
31	a:62bis	When she came ben she bobbit, e	vn, vc, pf	1803	T v, 220	by Neukomm
32	a:22bis	White cockade, The, D	vn, vc, pf	1803	T iv, 188	by Neukomm

BIBLIOGRAPHY

A: Basic biographies, collected letters, bibliographies. B: Catalogues, sources, research. C: Specialist publications, commemorative issues. D: Biography. E: Life: particular aspects. F: Publishers. G: Iconography. H: Authenticity. I: Works: general. J: Sacred vocal music. K: Operas. L: Secular vocal. M: Orchestral. N: Chamber without keyboard. O: Keyboard. P: Reputation.

A: Basic biographies, collected letters, bibliographies

G.A. Griesinger: 'Biographische Notizen über Joseph Haydn', *AMZ*, xi (Leipzig, 1808–9), 641–9, 657–68, 673–81, 689–99, 705–13, 721–33, 737 47, 776–81; also pubd separately (Leipzig, 1810/*R*, 2/1819; Eng. trans. in Gotwals, 1963); ed. F. Grasberger (Vienna, 1954); ed. K.H. Köhler (Leipzig, 1975); ed. P. Krause (Leipzig, 1979)

A.C. Dies: *Biographische Nachrichten von Joseph Haydn* (Vienna, 1810; Eng. trans. in Gotwals, 1963); ed. H. Seeger (Berlin, 1959, 4/1976)

C.F. Pohl *Joseph Haydn*, i (Berlin, 1875, 2/1878/*R*), ii (Leipzig, 1882/*R*), iii (Leipzig, 1927/*R*) [completed by H. Botstiber]

H.C.R. Landon, ed.: *The Collected Correspondence and London Notebooks of Joseph Haydn* (London, 1959)

V. Gotwals, ed.: *Joseph Haydn: Eighteenth-Century Gentleman and Genius* (Madison, WI, 1963, 2/1968 as *Haydn: Two Contemporary Portraits*) [trans. of Griesinger (1810) and Dies (1810)]

D. Bartha, ed.: *Joseph Haydn: gesammelte Briefe und Aufzeichnungen: unter Benützung der Quellensammlung von H.C. Robbins Landon* (Kassel, 1965)

A.P. Brown, J.T. Berkenstock and C.V. Brown: 'Joseph Haydn in Literature: a Bibliography', *Haydn-Studien*, iii/3 4 (1974), 173–352

H.C.R. Landon: *Haydn: Chronicle and Works*, i: *Haydn: the Early Years 1732 1765* (London, 1980); ii: *Haydn at Eszterháza 1766 1790* (1978); iii: *Haydn in England 1791–1795* (1976), iv: *Haydn: the Years of 'The Creation' 1796–1800* (1977); v: *Haydn: the Late Years 1801–1809* (1977)

H.C.R. Landon: *Haydn: a Documentary Study* (London, 1981)

H. Walter: 'Haydn-Bibliographie 1973–1983', *Haydn-Studien*, v/4 (1985), 205–93; 'Haydn-Bibliographie 1984–1990', vi/3 (1992), 173 238

M. Vignal: *Joseph Haydn* (Paris, 1988)

B: Catalogues, sources, research (see also E)

A. Fuchs: *Thematisches Verzeichniss der sämmtlichen Compositionen von Joseph Haydn zusammengestellt ...1839*, ed. R. Schaal (Wilhelmshaven, 1968)

A. Artaria: *Verzeichnis von musikalischen Autographen, revidierten Abschriften und einigen seltenen gedruckten Original-Ausgaben* (Vienna, 1893)

A. Orel: *Katalog der Haydn-Gedächtnisausstellung Wien 1932* (Vienna, 1932)

J.P. Larsen: *Die Haydn-Überlieferung* (Copenhagen, 1939, 2/1980); portions trans. in *Handel, Haydn, and the Viennese Classical Style* (Ann Arbor, 1988)

J.P. Larsen, ed.: *Drei Haydn Kataloge in Faksimile mit Einleitung und ergänzenden Themenverzeichnissen* (Copenhagen, 1941; Eng. trans., rev., 1989)

A. van Hoboken: *Joseph Haydn: thematisch-bibliographisches Werkverzeichnis*, i: *Instrumentalwerke*; ii: *Vokalwerke*; iii: *Register: Addenda und Corrigenda* (Mainz, 1957–78)

H.C.R. Landon: 'Survey of the Haydn Sources in Czechoslovakia", *Konferenz zum Andenken Joseph Haydns: Budapest 1959* (Budapest, 1961), 69–77

L. Nowak, ed.: *Joseph Haydn: Ausstellung zum 150. Todestag: vom 29. Mai bis 30. September 1959* (Vienna, 1959)

J. Vécsey and others, eds.: *Haydn művei az Országos Széchényi Könyvtár zenei gyűjteményében: kiadásra került az 1809–1959 évforduló alkalmából* [Haydn compositions in the National Széchényi Library, Budapest: published on the 150th anniversary of Haydn's death] (Budapest, 1959; Ger. trans., 1959; Eng. trans., 1960)

G. Feder: 'Zur Datierung Haydnscher Werke', *Anthony van Hoboken: Festschrift*, ed. J. Schmidt-Görg (Mainz, 1962), 50–54

G. Feder: 'Die Überlieferung und Verbreitung der handschriftlichen Quellen zu Haydns Werken (Erste Folge)', *Haydn-Studien*, i/1 (1965), 3–42; Eng. trans., *Haydn Yearbook 1968*, 102–39

G. Feder: 'Bemerkungen zu Haydns Skizzen', *BeJb 1973–7*, 69–86

I. Becker-Glauch: 'Haydn, Franz Joseph', *Einzeldrucke vor 1800*, RISM, A/I/4 (1974), 140–279

G. Feder: 'The Collected Works of Joseph Haydn', *Haydn Studies: Washington DC 1975* (New York, 1981), 26–34

J.P. Larsen: 'A Survey of the Development of Haydn Research', ibid., 14–25

S.C. Fisher: 'A Group of Haydn Copies for the Court of Spain', *Haydn-Studien*, iv/2 (1978), 65–84

G. Feder: 'Joseph Haydns Skizzen und Entwürfe: Übersicht der Manuskripte, Werkregister, Literatur- und Ausgabenverzeichnis', *FAM*, xxvi (1979), 172–88

G. Feder: 'Über Haydns Skizzen zu nicht identifizierten Werken', *Ars musica – musica scientica: Festschrift Heinrich Hüschen*, ed. D. Altenburg (Cologne, 1980), 100–11

E. Radant: 'A Facsimile of Hummel's Catalogue of the Princely Music Library in Eisenstadt, with Transliteration and Commentary', *Haydn Yearbook 1980*, 5–182

S.C. Bryant and G.W. Chapman: *A Melodic Index to Haydn's Instrumental Music: a Thematic Locator ...* (New York, 1982)

G. Mraz, G. Mraz and G. Schlag, eds.: *Joseph Haydn in seiner Zeit* (Eisenstadt, 1982) [exhibition catalogue]

E. Radant: 'A Thematic Catalogue of the Esterházy Archives (c1801–5)', *Haydn Yearbook 1982*, 180–212

A. Tyson: 'Paper Studies and Haydn: What Needs to be Done', *Joseph Haydn: Vienna 1982* (Munich, 1986), 577–92

G. Feder: 'Textkritische Methoden: Versuch eines Überblicks mit Bezug auf die Haydn-Gesamtausgabe', *Haydn-Studien*, v/2 (1983), 77–109

Katalog der Sammlung Anthony van Hoboken in der Musiksammlung der Österreichischen Nationalbibliothek, vi: *Joseph Haydn: Symphonien* (Tutzing, 1987); vii: *Joseph Haydn: Instrumentalmusik (Hob. II bis XI)* (1988); viii: *Joseph Haydn: Instrumentalmusik (Hob. XIV–XX/1)* (1990); ix: *Joseph Haydn: Vokalmusik* (1991)

H.A. Schafer: 'A Wisely Ordered Phantasie': Joseph Haydn's Creative Process from Sketches and Drafts for Instrumental Music (diss., Brandeis U., 1987)

R. von Zahn: 'Der fürstlich Esterházysche Notenkopist Joseph Elssler sen.', Haydn-Studien, vi/2 (1988), 130–47

Fine Music Manuscripts, Sotheby's, 21 Nov 1990 (London, 1990), lot 88 [sale catalogue]

Joseph und Michael Haydn: Autographe und Abschriften: Katalog (Munich, 1990) [pubn of the Musikabteilung, Staatsbibliothek Preussischer Kulturbesitz, Berlin]

F.K. Grave and M.G. Grave: Franz Joseph Haydn: a Guide to Research (New York, 1990)

D. Link: 'Vienna's Private Theatrical and Musical Life, 1783–92, as Reported by Count Karl Zinzendorf', JRMA, cxxii (1997), 205–57

S. Gerlach: 'Haydn's "Entwurf-Katalog": Two Questions', Haydn, Mozart, & Beethoven: Essays in Honour of Alan Tyson, ed. S. Brandenburg (Oxford, 1998), 65–83

C: Specialist publications, commemorative issues

Die Musik, viii/3 [no.16] (1908–9)

IMusSCR III: Vienna 1909 (Vienna, 1909)

Die Musik, xxiv/6 (1931–2)

Burgenländische Heimatblätter, i/1 (Eisenstadt, 1932)

MQ, xviii/2 (1932)

MT, lxxiii/March (1932)

ZfM, Jg.99 (1932), no.4

Burgenländische Heimatblätter, xxi/2 (Eisenstadt, 1959)

Konferenz zum Andenken Joseph Haydns: Budapest 1959 (Budapest, 1961)

ÖMz, xiv/5–6 (1959)

ZT, viii (1960) [Haydn memorial issue, with Ger. summaries]

Haydn Yearbook 1962–

Haydn-Studien, i– (1965–)

Der junge Haydn: Graz 1970 (Graz, 1972)

Jb für österreichische Kulturgeschichte, ii: Joseph Haydn und seine Zeit (1972)

Haydn Studies: Washington DC 1975 (New York, 1981)

Joseph Haydn und die Literatur seiner Zeit, ed. H. Zeman (Eisenstadt, 1976)

Haydn e il suo tempo: Siena 1979 [Chigiana, xxxvi, new ser. xvi (1979)]

Joseph Haydn a hudba jeho doby: Bratislava 1982 [Haydn and the music of his time] (Bratislava, 1984)

Joseph Haydn: Cologne 1982 (Regensburg, 1985)

Joseph Haydn: Vienna 1982 (Munich, 1986)

MQ, lxviii/4 (1982)

MT, cxxiii/March (1982)

ÖMz, xxxvii/3–4 (1982)

Musik-Konzepte, no.41 (1985)

J.P. Larsen: Handel, Haydn, and the Viennese Classical Style (Ann Arbor, 1988)

Haydn and his World, ed. E. Sisman (Princeton, NJ, 1997)

Haydn Studies, ed. W.D. Sutcliffe (Cambridge, 1998)

D: Biography

GerberNL

Framery: *Notice sur Joseph Haydn* (Paris, 1810)

G. Carpani: *Le Haydine, ovvero Lettere su la vita e le opere del celebre maestro Giuseppe Haydn* (Milan, 1812, 2/1823/*R*; Eng. trans., 1839 as *The Life of Haydn in Letters*)

Stendhal: *Lettres écrites de Vienne en Autriche, sur le célèbre compositeur Joseph Haydn, suivies d'une vie de Mozart, et de considérations sur Métastase et l'état présent de la musique en France et en Italie* (Paris, 1814, 3/1872, rev. 1928 by H. Martineau as *Vies de Haydn, de Mozart et de Métastase*; Eng. trans. R. Brewin, 1817, 2/1818, rev. 1972 by R.N. Coe as *Lives of Haydn, Mozart and Metastasio*) [plagiarism of Carpani (1812)]

C. von Wurzbach: *Joseph Haydn und sein Bruder Michael: zwei biobibliographische Künstler-Skizzen* (Vienna, 1861)

M. Brenet: *Haydn* (Paris, 1909, 2/1910; Eng. trans., 1926/*R*)

A. Schnerich: *Joseph Haydn und seine Sendung* (Zürich, 1922, 2/1926 with suppl. by W. Fischer)

K. Geiringer: *Joseph Haydn* (Potsdam, 1932, 2/1959)

K. Geiringer: *Haydn: a Creative Life in Music* (New York, 1946, enlarged 3/1982)

R. Hughes: *Haydn* (London, 1950, 6/1989)

L. Nowak: *Joseph Haydn: Leben, Bedeutung und Werk* (Zürich, 1951, 3/1966)

J. Webster: 'Prospects for Haydn Biography after Landon', *MQ*, lxviii (1982), 476–95

M. Huss: *Joseph Haydn: Klassiker zwischen Barock und Biedermeier* (Eisenstadt, 1984)

H.C.R. Landon and D. Wyn Jones: *Haydn: his Life and Music* (London, 1988)

W. Marggraf: *Joseph Haydn: Versuch einer Annäherung* (Leipzig, 1990)

E. Life: particular aspects

Ancestry, early years, last years, acquaintances, character

R. Bernhardt: 'Aus der Umwelt der Wiener Klassiker: Freiherr Gottfried van Swieten (1734–1803)', *Der Bär: Jb von Breitkopf & Härtel 1929–30*, 74–166

R.F. Müller: 'Heiratsbrief, Testament und Hinterlassenschaft der Gattin Joseph Haydns', *Die Musik*, xxii (1929–30), 93–9

H. Botstiber: 'Haydn and Luigia Polzelli', *MQ*, xviii (1932), 208–15

E.F. Schmid: *Joseph Haydn: ein Buch von Vorfahren und Heimat des Meisters* (Kassel, 1934)

E.F. Schmid: 'Joseph Haydns Jugendliebe', *Festschrift Wilhelm Fischer*, ed. H. von Zingerle (Innsbruck, 1956), 109–22

O.E. Deutsch: 'Haydn als Sammler', *ÖMz*, xiv (1959), 188–93

J.P. Larsen: 'Haydn und Mozart', *ÖMz*, xiv (1959), 216–22; Eng. trans. in Larsen, C1988

E.H. Müller von Asow: 'Joseph Haydns Tod in zeitgenössischen Berichten', *Musikerziehung*, xii (1959), 141–7

E. Schenk: 'Das Weltbild Joseph Haydns', *Österreichische Akademie der Wissenschaften: Almanach*, cix (1959), 245–72; repr. in *Ausgewählte Aufsätze, Reden und Vorträge* (Graz, 1967), 86–99

V. Gotwals: 'Joseph Haydn's Last Will and Testament', *MQ*, xlvii (1961), 331–53

E. Olleson: 'Gottfried van Swieten, Patron of Haydn and Mozart', *PRMA*, lxxxix (1962–3), 63–74

F. Blume: 'Haydn als Briefschreiber', *Syntagma musicologicum: gesammelte Reden und Schriften* (Kassel, 1963), 564–70

L. Somfai: 'Haydns Tribut an seinen Vorganger Werner', *Haydn Yearbook 1963–4*, 75–80

E. Olleson: 'Georg August Greisinger's Correspondence with Breitkopf and Härtel', *Haydn Yearbook 1965*, 5–53

G. Thomas: 'Greisinger's Briefe über Haydn: aus seiner Korrespondenz mit Breitkopf and Härtel', *Haydn-Studien*, i/2 (1966), 49–114

H. Walter: 'Die biographischen Beziehungen zwischen Haydn und Beethoven', *GfMKB: Bonn 1970* (Kassell, 1972), 79–83

G. Feder: 'Joseph Haydn als Mensch und Musiker', *Jb für österreichische Kulturgeschichte*, ii: *Joseph Haydn und seine Zeit* (1972), 43–56; also in *ÖMz*, xxvii (1972), 57–68

R. Angermüller: 'Sigismund Ritter von Neukomm (1778–1858) und seine Lehrer Michael und Joseph Haydn: eine Dokumentation', *Haydn-Studien*, iii/1 (1973), 29–42

R. Angermüller: 'Neukomms schottlische Liedbearbeitungen für Joseph Haydn', *Haydn-Studien*, iii/2 (1974), 151–3

K. Geiringer: 'Haydn and his Viennese Background', *Haydn Studies: Washington DC 1975* (New York, 1981), 3–13

M. Hörwarthner: 'Joseph Haydns Bibliotek: Versuch einer literaturhistorischen Rekonstrucktion', *Joseph Haydn und die Literatur seiner Zeit*, ed. H. Zeman (Eisenstadt, 1976), 157 207; Eng. trans. in *Haydn and his World*, ed. E. Sisman (Princeton NJ, 1997), 395–462

O. Biba: 'Nachrichten zur Musikpflege in der gräflichen Familie Harrach', *Haydn Yearbook 1978*, 36–44

D. Heartz: 'Haydn und Gluck im Burgtheater um 1760: Der neue krumme Teufel, Le Diable à quatre und die Sinfonie "Le soir"', *GfMKB: Bayreuth 1981* (Kassel, 1984), 120–35

R.N. Freeman: 'Robert Kimmerling: a Little-Known Haydn Pupil', *Haydn Yearbook 1982*, 143 79

G. Gruber: 'Doppelgesichtiger Haydn?', *ÖMz*, xxxvii (1982), 139 46

H. Walter: 'Kalkbrenners Lehrjahre und sein Unterricht bei Haydn', *Haydn-Studien*, v/1 (1982), 23–41

J. Webster: 'The Falling-Out between Haydn and Beethoven: the Evidence of the Sources', *Beethoven Essays: Studies in Honor of Elliot Forbes*, ed. L. Lockwood and P. Benjamin (Cambridge, MA, 1984), 3–45

J. Hurwitz: 'Haydn and the Freemasons', *Haydn Yearbook 1985*, 5–98

O. Biba: *'Eben komme ich von Haydn': Georg August Griesingers Korrespondenz mit Joseph Haydns Verleger Breitkopf & Härtel 1799–1819* (Zürich, 1987)

O. Biba: 'Haydns Kirchenmusikdienste für Graf Haugwitz', *Haydn-Studien*, vi/4 (1994), 278–87

D. Edge: 'New Sources for Haydn's Early Biography' [unpubd paper read at AMS 1993]

O. Biba: 'Haydns Kirchenmusikdienste für Graf Haugwitz', *Haydn-Studien*, vi/4 (1994), 278–87

Eisenstadt, Eszterháza, England, travels

T.G. von Karajan: *Joseph Haydn in London 1791 und 1792* (Vienna, 1861/R); also in *Jb für Vaterländische Geschichte* (1861)

C.F. Pohl: *Mozart und Haydn in London*, ii: *Haydn in London* (Vienna, 1867/R)

A. Csatkai: 'Die Beziehungen Gregor Josef Werners, Joseph Haydns und der fürstlichen Musiker zur Eisenstädter Pfarrkirche', *Burgenländische Heimatblätter*, i (Eisenstadt, 1932), 13–17

E.F. Schmid: 'Joseph Haydn in Eisenstadt: ein Beitrag zur Biographie des Meisters', *Burgenländische Heimatblätter*, i (Eisenstadt, 1932), 2–13

A. Valkó: 'Haydn magyarországi működése a levéltári akták tükrében' [Haydn's activities in Hungary, as revealed in the archives], *ZT*, vi (1957), 627–67; viii (1960), 527–668 [with Ger. summaries]

J. Harich: *Esterházy-Musikgeschichte im Spiegel der zeitgenössischen Textbücher* (Eisenstadt, 1959) [orig. in *H-Bn*]

M. Horányi: *Eszterházi vigasságok* (Budapest, 1959; Eng. trans., 1962, as *The Magnificence of Eszterháza*)

D. Bartha and L. Somfai: *Haydn als Opernkapellmeister: die Haydn-Dokumente der Esterházy-Opernsammlung* (Budapest, 1960)

J. Harich: 'Das Repertoire des Opernkapellmeisters Joseph Haydn in Eszterháza (1780–1790)', *Haydn Yearbook 1962*, 9–110

J. Harich: 'Haydn Documenta', *Haydn Yearbook 1963–4*, 2–44; *1965*, 122–52; *1968*, 39–101; *1970*, 47–168; *1971*, 70–163

B. Matthews: 'Haydn's Visit to Hampshire and the Isle of Wight, Described from Contemporary Sources', *Haydn Yearbook 1965*, 111–21

J. Harich: 'Das fürstlich Esterházy'sche Fideikommiss', *Haydn Yearbook 1968*, 5–38

G. Feder: 'Haydn und Eisenstadt', *ÖMz*, xxv (1970), 213–21

J. Harich: 'Das Opernensemble zu Eszterháza im Jahr 1780', *Haydn Yearbook 1970*, 5–46

J. Harich: 'Das Haydn-Orchester im Jahr 1780', *Haydn Yearbook 1971*, 5–69

P. Bryan: 'Haydn's Hornists', *Haydn-Studien*, iii/1 (1973), 52–8

J. Harich: 'Inventare der Esterházy-Hofmusikkapelle in Eisenstadt', *Haydn Yearbook 1975*, 5–125

S. Gerlach: 'Haydns Orchestermusiker von 1761 bis 1774', *Haydn-Studien*, iv/1 (1976), 35–48

U. Tank: 'Die Dokumente der Esterházy-Archive zur fürstlichen Hofkapelle in der Zeit von 1761 bis 1770', *Haydn-Studien*, iv/3–4 (1980), 129–333

G. Thomas: 'Haydn-Anekdoten', *Ars musica, musica scientia: Festschrift Heinrich Hüschen*, ed. D. Altenburg (Cologne, 1980), 435–43

U. Tank: *Studien zur Esterházyschen Hofmusik von etwa 1620 bis 1790* (Regensburg, 1981)

'The *Acta Musicalia* of the Esterházy Archives', *Haydn Yearbook 1982*, 5–96; *1983*, 9–128; *1984*, 93–180; *1985*, 99–207; *1992*, 1–84; *1993*, 115–96

R. Hellyer: 'The Wind Ensembles of the Esterházy Princes', *Haydn Yearbook 1984*, 5–92

'Documents from the Archives of János Harich', *Haydn Yearbook 1993*, 1–109; *1994*, 1–359

Miscellaneous

G. Nottebohm: *Beethoven's Studien*, i: *Beethoven's Unterricht bei J. Haydn, Albrechtsberger und Salieri: nach den Original-Manuskripten dargestellt* (Leipzig and Winterthur, 1873/R)

F. von Reinöhl: 'Neues zu Beethovens Lehrjahr bei Haydn', *NBeJb 1935*, 36–47

N.A. Solar-Quintes: 'Las relaciones de Haydn con la casa de Benavente'. *AnM*, ii (1947), 81–8

O.E. Deutsch: 'Haydn bleibt Lehrling: nach den Freimaurer-Akten des Österreichischen Staatsarchivs', *Musica*, xiii (1959), 289–90

H. Seeger: 'Zur musikhistorischen Bedeutung der Haydn-Biographie von Albert Christoph Dies (1810)', *BMw*, i/3 (1959), 24–37 [incl. Neukomm's remarks about Dies (1810)]

E. Olleson: 'Haydn in the Diaries of Count Karl von Zinzendorf', *Haydn Yearbook 1963–4*, 45–63

H. Unverricht: 'Unveröffentlichte und wenig bekannte Briefe Joseph Haydns', *Mf*, xviii (1965), 40–45

E. Radant, ed.: 'Die Tagebücher von Joseph Carl Rosenbaum 1770–1829', *Haydn Yearbook 1968*, 7–159

C.-G. Stellan Mörner: 'Haydniana aus Schweden um 1800', *Haydn-Studien*, ii/1 (1969), 1–33

J. Chailley: 'Joseph Haydn and the Freemasons', *Studies in Eighteenth-Century Music: a Tribute to Karl Geiringer*, ed. H.C.R. Landon and R.E. Chapman (New York and London, 1970), 117–24

A. Mann: 'Beethoven's Contrapuntal Studies with Haydn', *MQ*, lvi (1970), 711–26; another version in *GfMKB: Bonn 1970* (Kassel, 1972), 70–74

A. Mann: 'Haydn as Student and Critic of Fux', *Studies in Eighteenth-Century Music: a Tribute to Karl Geiringer*, ed. H.C.R. Landon and R.E. Chapman (New York and London, 1970), 323–32; another version in *Musik und Verlag: Karl Vötterle zum 65. Geburtstag*, ed. R. Baum and W. Rehm (Kassel, 1968), 433–7

M. Pandi and H. Schmidt: 'Musik zur Zeit Haydns und Beethovens in der Pressburger Zeitung', *Haydn Yearbook 1971*, 165–293

A. Mann: 'Haydn's Elementarbuch: a Document of Classic Counterpoint Instruction', *Music Forum*, iii (1973), 197–237

G. Feder and S. Gerlach: 'Haydn-Dokumente aus dem Esterházy-Archiv in Forchtenstein', *Haydn-Studien*, iii/2 (1974), 92–105

H. Zeman, ed.: *Joseph Haydn und die Literatur seiner Zeit* (Eisenstadt, 1976)

F: Publishers

W. Sandys and S.A. Forster: *The History of the Violin, and Other Instruments* (London, 1864) [correspondence of Haydn and Forster]

F. Artaria and H. Botstiber: *Joseph Haydn und das Verlagshaus Artaria: nach den Briefen des Meisters an das Haus Artaria & Compagnie dargestellt* (Vienna, 1909)

H. von Hase: *Joseph Haydn und Brietkopf & Härtel* (Leipzig, 1909)

G. Feder: 'Die Eingriffe des Musikverlegers Hummel in Haydns Werken', *Musicae scientiae collectanea: Festschrift Karl Gustav Fellerer zum siebzigsten Geburtstag*, ed. H. Hüschen (Cologne, 1973), 88–101

H.E. Poole: 'Music Engraving Practice in Eighteenth-Century London: a Study of Some Forster Editions of Haydn and their Manuscript Sources', *Music and Bibliography: Essays in Honour of Alec Hyatt King*, ed. O. Neighbour (London, 1980), 98–131

N.A. Mace: 'Haydn and the London Music Sellers: Forster and Longman & Broderip', *ML*, lxxvii (1996), 527–41

G: Iconography

A. Fuchs: 'Verzeichniss aller Abbildungen Joseph Haydn's', *Wiener Allgemeine Musik-Zeitung*, vi (1846), 237–9

J. Muller: 'Haydn Portraits', *MQ*, xviii (1932), 282–98

L. Somfai: *Joseph Haydn: sein Leben in zeitgenössischen Bildern* (Budapest and Kassel, 1966; Eng. trans., 1969)

J.P. Larsen: 'Zur Frage der Porträtähnlichkeit der Haydn-Bildnisse', *SMH*, xii (1970), 153–66

H: Authenticity (see also B)

J.P. Larsen: 'Haydn und das "kleine Quartbuch"', *AcM*, vii (1935), 111–23; Eng. trans. in Larsen, 1988 [beginning of Larsen–Sandberger controversy; see Brown, Berkenstock and Brown (1974), no.1134]

H.C.R. Landon: 'Problems of Authenticity in Eighteenth-Century Music', *Instrumental Music: Cambridge, MA, 1957* (Cambridge MA, 1959), 31–56

E. Schenk: 'Ist die Göttweiger Rorate-Messe ein Werk Joseph Haydns?', *SMw*, xxiv (1960), 87–105

A. Tyson: 'Haydn and Two Stolen Trios', *MR*, xxii (1961), 21–7

P. Mies: 'Anfrage zu einem Jos. Haydn unterschobenen Werk', *Haydn Yearbook 1962*, 200–01

A. Tyson and H.C.R. Landon: 'Who Composed Haydn's Op.3?', *MT*, cv (1964), 506–7

H. Schwarting: 'Über die Echtheit dreier Haydn-Trios', *AMw*, xxii (1965), 169–82

L. Somfai: 'Zur Echtheitsfrage des Haydn'schen "Opus 3"', *Haydn Yearbook 1965*, 153–65

H. Unverricht, A. Gottron and A. Tyson: *Die beiden Hoffstetter: zwei Komponisten-Porträts mit Werkverzeichnissen* (Mainz, 1968)

J.P. Larsen: 'Über die Möglichkeiten einer musikalischen Echtheitsbestimmung für Werke aus der Zeit Mozarts und Haydns', *MJb 1971–2*, 7–18; Eng. trans. in Larsen, 1988

G. Feder: 'Die Bedeutung der Assoziation und des Wertvergleichs für das Urteil in Echtheitsfragen', *IMSCR XI: Copenhagen 1972* (Copenhagen, 1974), i, 365–77

C.E. Hatting: 'Haydn oder Kayser?: eine Echtheitsfrage', *Mf*, xxv (1972), 182–7

W.T. Marrocco: 'The String Quartet attributed to Benjamin Franklin', *Proceedings of the American Philosophical Society*, cxvi (1972), 477–85

G. Feder: 'Apokryphe "Haydn"-Streichquartette', *Haydn-Studien*, iii/2 (1974), 125–50

D.L. Brantley: *Disputed Authorship of Musical Works: a Quantitative Approach to the Attribution of the Quartets Published as Haydn's Opus 3* (diss., U. of Iowa, 1978)

B.S. Brook: 'Determining Authenticity Through Internal Analysis: a Multifaceted Approach (with Special Reference to Haydn's String Trios)', *Joseph Haydn: Vienna 1982* (Munich, 1986), 551–66

B.C. MacIntyre: 'Haydn's Doubtful and Spurious Masses: an Attribution Update', *Haydn-Studien*, v/1 (1982), 42–54

J. Spitzer: *Authorship and Attribution in Western Art Music* (diss., Cornell U., 1983) [chap.4, Haydn]

S. Fruehwald: *Authenticity Problems in Franz Joseph Haydn's Early Instrumental Works: a Stylistic Investigation* (New York, 1988)

I: Works: general

W.H. Hadow: *A Croatian Composer: Notes towards the Study of Joseph Haydn* (London, 1897); repr. in *Collected Essays* (London, 1928)

T. de Wyzewa: 'A propos du centenaire de la mort de Joseph Haydn', *Revue des deux mondes*, 5th period, li (1909), 935–46

H. Jalowetz: 'Beethovens Jugendwerke in ihren melodischen Beziehungen zu Mozart, Haydn und Ph.E. Bach', *SIMG*, xii (1910–11), 417–74

W. Fischer: 'Zur Entwicklungsgeschichte des Wiener klassischen Stils', *SMw*, iii (1915), 24–84

H. Kretzschmar: *Führer durch den Konzertsaal* (Leipzig, 5/1919–21)

G. Becking: *Studien zu Beethovens Personalstil: das Scherzothema: mit einem bisher unveröffentlichten Scherzo Beethovens* (Leipzig, 1921) [chap.2, Haydn's minuets]

F. Blume: 'Fortspinnung und Entwicklung: ein Beitrag zur musikalischen Begriffsbildung', *JbMP 1929*, 51–70; repr. in *Syntagma musicologicum: gesammelte Reden und Schriften* (Kassel, 1963), 504–25

E.F. Schmid: 'Joseph Haydn und Carl Philipp Emanuel Bach', *ZMw*, xiv (1931–2), 299–312

G. Adler: 'Haydn and the Viennese Classical School', *MQ*, xviii (1932), 191–207

R. von Tobel: *Die Formenwelt der klassischen Instrumentalmusik* (Berne, 1935)

D.F. Tovey: *Essays in Musical Analysis* (London, 1935–44/R)

T. Georgiades: 'Zur Musiksprache der Wiener Klassiker', *MJb 1951*, 50–60

J.P. Larsen: 'Zu Haydns künstlerischer Entwicklung', *Festschrift Wilhelm Fischer*, ed. H. von Zingerle (Innsbruck, 1956), 123–9; Eng. trans in Larsen, 1988

E.F. Schmid: 'Mozart and Haydn', *MQ*, xlii (1956), 145–61

F. Noske: 'Le principe structural génétique dans l'oeuvre instrumental de Joseph Haydn', *RBM*, xii (1958), 35–9

H. Besseler: 'Einflüsse der Contratanzmusik auf Joseph Haydn', *Konferenz zum Andenken Joseph Haydns: Budapest 1959* (Budapest, 1961), 25–40

H. Engel: 'Haydn, Mozart und die Klassik', *MJb 1959*, 46–79; another version in *IMSCR VIII: New York 1961* (Kassel, 1961–2), i, 285–304

B. Szabolcsi: 'Joseph Haydn und die ungarische Musik', *BMw*, i/2 (1959), 62–73; repr. in *Konferenz zum Andenken Joseph Haydns: Budapest 1959* (Budapest, 1961), 159–75

H. Schwarting: 'Ungewöhnliche Repriseneintritte in Haydns späterer Instrumentalmusik', *AMw*, xvii (1960), 168–82

G. Feder: 'Bemerkungen über die Ausbildung der klassischen Tonsprache in der Instrumentalmusik Haydns', *IMSCR VIII: New York 1961* (Kassel, 1961–2), i, 305–13

G. Feder: 'Eine Methode der Stiluntersuchung, demonstriert an Haydns Werken', *GfMKB: Leipzig 1966* (Kassel, 1970), 275–85

D. Bartha: 'Volkstanz-Stilisierung in Joseph Haydns Finale-Themen', *Festschrift für Walter Wiora*, ed. L. Finscher and C.-H. Mahling (Kassel, 1967), 375–84

L. Schrade: 'Joseph Haydn als Schöpfer der klassischen Musik', *De scientia musicae studia atque orationes*, ed. E. Lichtenhahn (Berne, 1967), 506–18

M.S. Cole: 'The Rondo Finale: Evidence for the Mozart–Haydn Exchange?', *MJb 1968–70*, 242–56

B.S. Brook: 'Sturm und Drang and the Romantic Period in Music', *Studies in Romanticism*, ix (1970), 269–84

G. Feder: 'Die beiden Pole im Instrumentalschaffen des jungen Haydn', *Der junge Haydn: Graz 1970* (Graz, 1972), 192–201

G. Feder: 'Similarities in the Works of Haydn', *Studies in Eighteenth-Century Music: a Tribute to Karl Geiringer*, ed. H.C.R. Landon and R.E. Chapman (New York and London, 1970), 186–97

G. Feder: 'Stilelemente Haydns in Beethovens Werken', *GfMKB: Bonn 1970* (Kassel, 1972), 65–70

G. Gruber: 'Musikalische Rhetorik und barocke Bildlichkeit in Kompositionen des jungen Haydn', *Der junge Haydn: Graz 1970* (Graz, 1972), 168–91

C. Rosen: *The Classical Style: Haydn, Mozart, Beethoven* (New York, 1971, enlarged 3/1997) [incl. CD]

D.S. Cushman: *Joseph Haydn's Melodic Materials: an Exploratory Introduction to the Primary and Secondary Sources together with an Analytical Catalogue and Tables of Proposed Melodic Correspondence and/or Variance* (diss., Boston U., 1972)

L. Somfai: 'Vom Barock zur Klassik: Umgestaltung der Proportionen und des Gleichgewichts in zyklischen Werken Joseph Haydns', *Jb für österreichische Kulturgeschichte*, ii: *Joseph Haydn und seine Zeit* (1972), 64–72

W. Steinbeck: *Das Menuett in der Instrumentalmusik Joseph Haydns* (Munich, 1973)

S. Wollenberg: 'Haydn's Baryton Trios and the "Gradus"', *ML*, liv (1973), 170–78

E. Badura-Skoda: 'The Influence of the Viennese Popular Comedy on Haydn and Mozart', *PRMA*, c (1973–4), 185–99

L. Somfai: 'The London Revision of Haydn's Instrumental Style', *PRMA*, c (1973–4), 159–74

G. Chew: 'The Night-Watchman's Song quoted by Haydn and its Implications', *Haydn-Studien*, iii/2 (1974), 106–24

A.P. Brown: 'Critical Years for Haydn's Instrumental Music: 1787–90', *MQ*, lxii (1976), 374–94

H. Walter: 'Das Posthornsignal bei Haydn und anderen Komponisten des 18. Jahrhunderts', *Haydn-Studien*, iv/1 (1976), 21–34

C. Rosen: *Sonata Forms* (New York, 1980, 2/1988)

R.L. Todd: 'Joseph Haydn and the *Sturm und Drang*: a Revaluation', *MR*, lxi (1980), 172–96

S.E. Paul: *Wit, Comedy and Humour in the Instrumental Music of Franz Joseph Haydn* (diss., U. of Cambridge, 1981)

T. Istvánffy: *All'Ongarese: Studien zur Rezeption ungarischer Musik bei Haydn, Mozart und Beethoven* (diss., U. of Heidelberg, 1982)

J. LaRue: 'Multistage Variance: Haydn's Legacy to Beethoven', *JM*, i (1982), 265–74

F. Neumann: 'Bemerkungen über Haydns Ornamentik', *Joseph Haydn: Vienna 1982* (Munich, 1986), 35–42

J. Webster: 'Binary Variants of Sonata Form in Early Haydn Instrumental Music', ibid., 127–35

H. Krones: 'Das "hohe Komische" bei Joseph Haydn', *ÖMz*, xxxviii (1983), 2–8

P. Gülke: 'Nahezu ein Kant der Musik', *Musik-Konzepte*, no.41 (1985), 67–73

S. Schmalzriedt: 'Charakter und Drama: zur historischen Analyse von Haydnschen und Beethovenschen Sonatensätzen', *AMw*, xlii (1985), 37–66

M.E. Bonds: *Haydn's False Recapitulations and the Perception of Sonata Form in the Eighteenth Century* (diss., Harvard U., 1988)

E. Haimo: 'Haydn's Altered Reprise', *JMT*, xxxii (1988), 335–51

M. Poštolka: *Mlad'y Joseph Haydn: Jeho V'yvoj ke Klasickému Slohu* [The young Haydn: towards the Classical style] (Prague, 1988)

C. Willner: 'Chromaticism and the Mediant in Four Late Haydn Works', *Theory and Practice*, xiii (1988), 79–114

H. Federhofer: 'Tonsatz und Instrumentation', *Musicologica austriaca*, ix (1989), 45–57

W.D. Sutcliffe: 'Haydn's Musical Personality', *MT*, cxx (1989), 341–4

A. Raab: *Funktionen des Unisono, dargestellt an den Streichquartetten und Messen von Joseph Haydn* (Frankfurt, 1990)

M.E. Bonds: 'Haydn, Laurence Sterne, and the Origins of Musical Irony', *JAMS*, xliv (1991), 57–91

J. Webster: *Haydn's 'Farewell' Symphony and the Idea of Classical Style: Through Composition and Cyclic Integration in his Instrumental Music* (Cambridge, 1991)

N.S. Josephson: 'Modulatory Patterns in Haydn's Late Development Sections', *Haydn Yearbook 1992*, 181–91

G.A. Wheelock: *Haydn's Ingenious Jesting with Art: Contexts of Musical Wit and Humor* (New York, 1992)

E.R. Sisman: *Haydn and the Classical Variation* (Cambridge, MA, 1993)

F.K. Grave: 'Metrical Dissonance in Haydn', *JM*, xiii (1995), 168–202

D. Heartz: *Haydn, Mozart and the Viennese School 1740–1780* (New York, 1995)

M. Spitzer: 'The Retransition as Sign: Listener-Oriented Approaches to Tonal Closure in Haydn's Sonata Form Movements', *JRMA*, cxxi (1996), 11–45

P.A. Hoyt: 'Haydn's New Incoherence', *Music Theory Spectrum*, xix (1997), 264–84

M. Hunter: 'Haydn's London Piano Trios and his Salomon String Quartets: Private vs. Public?', *Haydn and his World*, ed. E. Sisman (Princeton, NJ, 1997), 103–30

E. Sisman: 'Haydn, Shakespeare, and the Rules of Originality', ibid., 3–56

A. Ballstaedt: '"Humor" und "Witz" in Joseph Haydns Musik', *AMw*, lv (1998), 195–219

D. Chua: 'Haydn as Romantic: a Chemical Experiment with Instrumental Music', *Haydn Studies*, ed. W. Sutcliffe (Cambridge, 1998), 120–51

G. Edwards: 'Papa Doc's Recap Caper: Haydn and Temporal Dyslexia', ibid., 291–320

M. Spitzer: 'Haydn's Reversals: Style Change, Gesture and the Implication-Realization Model', ibid., 177–217

L. Finscher: *Joseph Haydn und seine Zeit* (Laaber, 2000)

J: Sacred vocal music

A. Schnerich: *Der Messen-Typus von Haydn bis Schubert* (Vienna, 1892)

A. Sandberger: 'Zur Entstehungsgeschichte von Haydns "Sieben Worten des Erlösers am Kreuze"', *JbMP 1903*, 47–59; repr. in Ausgewählte Aufsätze zur Musikgeschichte, i (Munich, 1921/R), 266–81

M. Friedlaender: 'Van Swieten und das Textbuch zu Haydns Jahreszeiten', *JbMP 1909*, 47–56

H. Schenker: 'Haydn: Die Schöpfung: die Vorstellung des Chaos', *Meisterwerk in der Musik*, ii (1926/R), 159–70; Eng. trans. (Cambridge, 1996), 97–105

K. Geiringer: 'Haydn's Sketches for "The Creation"', *MQ*, xviii (1932), 299–308

C.M. Brand: *Die Messen von Joseph Haydn* (Würzburg, 1941/R)

K. Geiringer: 'The Small Sacred Works by Haydn in the Esterházy Archives at Eisenstadt', *MQ*, xlv (1959), 460–72

E.F. Schmid: 'Haydns Oratorium "Il ritorno di Tobia": seine Entstehung und seine Schicksale', *AMw*, xvi (1959), 292–313

M. Stern: 'Haydns "Schöpfung": Geist und Herkunft des van Swietenschen Librettos: ein Beitrag zum Thema "Säkularisation" im Zeitalter der Aufklärung', *Haydn-Studien*, i/3 (1966), 121–98

D. McCaldin: 'Haydn's First and Last Work: the "Missa Brevis" in F major', *MR*, xxviii (1967), 165–72

A. Riedel-Martiny: 'Das Verhältnis von Text und Musik in Haydns Oratorien', *Haydn-Studien*, i/4 (1967), 205–40

H. Walter: 'Gottfried van Swietens handschriftliche Textbücher zu "Schöpfung" und "Jahreszeiten"', *Haydn-Studien*, i/4 (1967), 241–77

E. Olleson: 'The Origin and Libretto of Haydn's "Creation"', *Haydn Yearbook 1968*, 148–68

I. Becker-Glauch: 'Neue Forschungen zu Haydns Kirchenmusik', *Haydn-Studien*, ii/3 (1970), 167–241 [with thematic catalogue of small sacred works]

M. Chusid: 'Some Observations on Liturgy, Text and Structure in Haydn's Late Masses', *Studies in Eighteenth-Century Music: a Tribute to Karl Geiringer*, ed. H.C.R. Landon and R.E. Chapman (New York and London, 1970), 125–35

A. Riethmüller: 'Die Vorstellung des Chaos in der Musik: zu Joseph Haydns Oratorium "Die Schöpfung"', *Convivium Cosmologicum ... Helmut Hönl zum 70. Geburtstag*, ed. A. Ginnaras (Basle, 1973), 185–95

H.-J. Horn: 'FIAT LVX: zum kunsttheoretischen Hintergrund der "Erschaffung" des Lichtes in Haydns Schöpfung', *Haydn-Studien*, iii/2 (1974), 65–84

J.T. Berkenstock: *The Smaller Sacred Compositions of Joseph Haydn* (diss., Northwestern U., 1975)

O. Moe: 'Structure in Haydn's The Seasons', *Haydn Yearbook 1975*, 340–48

J. Dack: *The Origins and Development of the Esterházy Kapelle in Eisenstadt until 1790* (diss., U. of Liverpool, 1976)

D. Heartz: 'The Hunting Chorus in Haydn's Jahreszeiten and the "Airs de chasse" in the Encyclopédie', *Journal of Eighteenth-Century Studies*, ix (1976), 523–39

J.P. Larsen: 'Beethoven C-dur Messe und die Spätmessen Joseph Haydns', *Beethoven Colloquium: Vienna 1977* (Kassel, 1978), 12–19; Eng. trans. in Larsen, 1988

O. Biba: 'Beispiele für die Besetzungsverhältnisse bei Aufführungen von Haydns Oratorien in Wien zwischen 1784 und 1808', *Haydn-Studien*, iv/2 (1978), 94–104

G. Feder: 'Haydns Korrekturen zum Klavierauszug der "Jahreszeiten"', *Festschrift Georg von Dadelsen*, ed. T. Kohlhase and V. Scherliess (Neuhausen-Stuttgart, 1978), 101–12

H. Zeman: 'Das Textbuch Gottfried van Swietens zu Joseph Haydns "Die Schöpfung"', *Die österreichische Literatur: ihr Profil an der Wende vom 18. zum 19. Jahrhundert*, ed. H. Zeman (Graz, 1979), 403–25

V. Ravizza: *Joseph Haydn: Die Schöpfung* (Munich, 1981)

H. Unverricht: 'Joseph Haydns *Die sieben Worte Christi am Kreuze* in der Bearbeitung des Passauer Hofkapellmeisters Joseph Friebert', *KJb*, lxv (1981), 83–94

B. Edelmann: 'Haydn's *Il ritorno di Tobia* und der Wandel des "Geschmacks" in Wien nach 1780', *Joseph Haydn: Cologne 1982* (Regensburg, 1985), 189–214

L.M. Kantner: 'Das Messenschaffen Joseph Haydns und seiner italienischen Zeitgenossen: ein Vergleich', ibid., 145–59

J.P. Larsen: 'Haydn's Early Masses: Evolution of a Genre', *American Choral Review*, xxiv/2–3 (1982), 48–60; repr. in Larsen, 1988

H.E. Smither: 'Haydn's *Il ritorno di Tobia* und die Tradition des italienischen Oratoriums', *Joseph Haydn: Cologne 1982* (Regensburg, 1985), 160–88

V. Kalisch: 'Haydn und die Kirchenmusik: ein analytischer Versuch am Beispiel des Benedictus der Schöpfungsmesse', *Musik und Kirche*, liv (1984), 159–70

W. Michel: 'Die Tobias-Dramen bis zu Haydns Oratorium "Il ritorno di Tobia"', *Haydn-Studien*, v/3 (1984), 147–68

H.C.R. Landon, ed.: *The Creation and The Seasons: the Complete Authentic Sources for the Word-Books* (Cardiff, 1985)

A.P. Brown: *Performing Haydn's 'The Creation': Reconstructing the Earliest Renditions* (Bloomington, IN, 1986)

G. Feder: 'Die *Jahreszeiten* in der Vertonung von Joseph Haydn', *Die vier Jahreszeiten im 18. Jahrhundert: Langenburg 1983* (Heidelberg, 1986), 96–107

T. Göllner: *Die Sieben Worte am Kreuz bei Schütz und Haydn* (Munich, 1986)

B.C. MacIntyre: *The Viennese Concerted Mass of the Early Classic Period* (Ann Arbor, 1986)

H. Zeman: 'Von der irdischen Glückseligkeit: Gottfried van Swietens *Jahreszeiten-Libretto*: eine Utopie vom natürlichen Leben des Menschen', *Die vier Jahreszeiten im 18. Jahrhundert: Langenburg 1983* (Heidelberg, 1986), 108–20

W. Kirsch: 'Vergangenes und Gegenwärtiges in Haydns Oratorien: zur Dramaturgie der "Schöpfung" und der "Jahreszeiten"', *Florilegium musicologicum: Hellmut Federhofer zum 75. Geburtstag*, ed. C.-H. Mahling (Tutzing, 1988), 169–87

F. Krummacher: 'Symphonische Verfahren in Haydns späten Messen', *Das musikalische Kunstwerk: Festschrift Carl Dahlhaus*, ed. H. Danuser and others (Laaber, 1988), 455–81

A.P. Brown: 'Haydn's Chaos: Genesis and Genre', *MQ*, lxxiii (1989), 18–59

D.W. Jones: 'Haydn's *Missa sunt bona mixta malis* and the *a cappella* Tradition', *Music in Eighteenth-Century Austria: Cardiff 1991* (Cambridge, 1996), 89–111

W.A. Kumbier: 'A "New Quickening": Haydn's *The Creation*, Wordsworth and the Pictorialist Imagination', *Studies in Romanticism*, xxx (1991), 535–63

N. Temperley: *Haydn: 'The Creation'* (Cambridge, 1991)

L. Kramer: 'Music and Representation: the Instance of Haydn's *Creation*', *Music and Text: Critical Inquiries*, ed. S.P. Sher (Cambridge, 1992), 139–62

L. Kramer: 'Haydn's Chaos, Schenker's Order; or, Hermeneutics and Musical Analysis: Can They Mix?', *19CM*, xvi (1992–3), 3–17

A.P. Brown: '*The Creation* and *The Seasons*: Some Allusions, Quotations, and Models from Handel to Mendelssohn', *CMc*, no.51 (1993), 26–58

W.A. Kumbier: 'Rhetoric in Haydn's Applausus', *Haydn Yearbook 1993*, 213–65

B.C. MacIntyre: 'Wiener Salve-Regina-Vertonungen und Haydn', *Haydn-Studien*, vi/4 (1994), 261–77

J. Webster: 'Haydns Salve Regina in g-Moll (1771) und die Entwicklung zum durch-komponierten Zyklus', *Haydn-Studien*, vi/4 (1994), 245–60

G. Chew: 'Haydn's Pastorellas: Genre, Dating and Transmission in the Early Church Works', *Studies in Music History Presented to H.C. Robbins Landon*, ed. O. Biba and D.W. Jones (London, 1996), 21–43

J. Webster: 'The *Creation*, Haydn's Late Vocal Music, and the Musical Sublime', *Haydn and his World*, ed. E. Sisman (Princeton, NJ, 1997), 57–102

B.C. MacIntyre: *Haydn: The Creation* (New York, 1998)

J.W. McGrann: 'Of Saints, Name Days, and Turks: Some Background on Haydn's Masses Written for Prince Nikolaus II Esterházy', *JMR*, xvii (1998), 195–210

J. Webster: 'Haydn's Sacred Vocal Music and the Aesthetics of Salvation', *Haydn Studies*, ed. W.D. Sutcliffe (Cambridge, 1998), 35–69

G. Feder: *Joseph Haydn: Die Schöpfung* (Kassel, 1999)

K: Operas

L. Wendschuh: *Über Joseph Haydn's Opern* (Halle, 1896)

R. Haas: 'Teutsche Comedie Arien', *ZMw*, iii (1920–21), 405–15

H. Wirth: *Joseph Haydn als Dramatiker: sein Bühnenschaffen als Beitrag zur Geschichte der deutschen Oper* (Wolfenbüttel, 1940)

H.C.R. Landon: 'Haydn's Marionette Operas and the Repertoire of the Marionette Theatre at Esterház Castle', *Haydn Yearbook 1962*, 111–99

G. Feder: 'Einige Thesen zu dem Thema: Haydn als Dramatiker', *Haydn-Studien*, ii/2 (1969), 126–30

G. Feder: 'Ein Kolloquium über Haydns Opern', *Haydn-Studien*, ii/2 (1969), 113–31 [incl. catalogue of roles in Haydn's operas]

E. Badura-Skoda: '"Teutsche Comoedie-Arien" und Joseph Haydn', *Der junge Haydn: Graz 1970* (Graz, 1972), 59–73

K. Geiringer: 'From Guglielmi to Haydn: the Transformation of an Opera', *IMSCR XI: Copenhagen 1972* (Copenhagen, 1974), i, 391–5

H.C.R. Landon: 'The Operas of Haydn', *NOHM*, vii (1973), 172–99

G. Feder: 'Opera seria, opera buffa und opera semiseria bei Haydn', *Opernstudien: Anna Amalie Abert zum 65. Geburtstag*, ed. K. Hortschansky (Tutzing, 1975), 37–55

A.P. Brown: 'Tommaso Traetta and the Genesis of a Haydn Aria (Hob. XXIVb:10)', *Haydn e il suo tempo: Siena 1979* [Chigiana, xxxvi, new ser. xvi (1979)], 101–42

L'avant-scène l'opéra, no.42 (1982) [*Orlando paladino* issue]

Joseph Haydn: Cologne 1982 (Regensburg, 1985) [incl. D. Altenburg: 'Haydn und die Tradition der italienischen Oper: Bemerkungen zum Opernrepertoire des Esterházyschen Hofes', 77–99; G. Allroggen: '*La canterina* in den Vertonungen von Nicolà Piccinni und Joseph Haydn', 100–12; F. Lippmann: 'Haydn und die opera buffa: Vergleiche mit italienischen Werken gleichen Textes', 113–40]

Joseph Haydn: Vienna 1982 (Munich, 1986) [incl. R. Steblin: 'Key Characteristics and Haydn's Operas', 91–100; E. Melkus: 'Haydn als Dramatiker am Beispiel der Oper *La vera costanza*', 256–76; R. Strohm: 'Zur Metrik in Haydns und Anfossis "La vera costanza"', 279–94; M. McClymonds: 'Haydn and his Contemporaries: *Armida abbandonata*', 325–32; D. Heartz: 'Haydn's *Acide e Galatea* and the Imperial Wedding Operas of 1760 by Hasse and Gluck', 332–40; S. Leopold: '*Le pescatrici*: Goldoni, Haydn, Gassmann', 341–9; G. Lazarevich: 'Haydn and the Italian Comic Intermezzo Tradition', 376–84; H. Geyer-Kiefl: 'Guglielmis *Le pazzie d'Orlando* und Haydns *Orlando Paladino*', 403–13]

F. Lippmann: 'Haydns "La fedeltà premiata" und Cimarosas "L'infedeltà fedele"', *Haydn-Studien*, v/1 (1982), 1–15

G. Feder: 'Haydn und Hasse', *Johann Adolf Hasse und die Musik seiner Zeit: Siena 1983* [*AnMc*, no.25 (1987)], 305–27

F. Lippmann: 'Haydns opere serie: Tendenzen und Affinitäten', *Studi musicali*, xii (1983), 301–31

M. Brago: 'Haydn, Goldoni and *Il mondo della luna*', *Journal of Eighteenth-Century Studies*, xvii (1984), 308–22

M. Hunter: 'Haydn's Sonata-Form Arias', *CMc*, nos.37–8 (1984), 19–32

J.A. Rice: 'Sarti's *Giulio Sabino*, Haydn's *Armida*, and the Arrival of Opera Seria at Eszterháza', *Haydn Yearbook 1984*, 181–98

G. Thomas: 'Haydns deutsche Singspiele', *Haydn-Studien*, vi/1 (1986), 1–63

B.A. Brown: '*Le pazzie d'Orlando*, *Orlando Paladino*, and the Uses of Parody', *Italica*, lxiv (1987), 583–605

Joseph Haydn und die Oper seiner Zeit: Eisenstadt 1988 (Eisenstadt, 1992)

P. Delby: 'Social Commentary in the Music of Haydn's Goldoni Operas', *Metaphor: a Musical Dimension: Melbourne 1988* (Sydney, 1991), 51–68

G. Feder and G. Thomas: 'Dokumente zur Ausstattung von *Lo speziale*, *L'infedeltà delusa*, *La fedeltà premiata*, *Armida* und anderen Opern Haydns', *Haydn-Studien*, vi/2 (1988), 88–115

M. Hunter: 'Text, Music and Drama in Haydn's Italian Opera Arias: Four Case Studies', *JM*, vii (1989), 29–57

C. Clark: 'Intertextual Play and Haydn's *La fedeltà premiata*', *CMc*, no.51 (1993), 59–81

R. Wochnik *Die Musiksprache in den opere semiserie Joseph Haydns unter besonderer Berücksichtigung von 'L'incontro improvviso*' (Hamburg, 1993)

R. Green: 'Representing the Aristocracy: the Operatic Haydn and *Le pescatrici*', *Haydn and his World*, ed. E. Sisman (Princeton, NJ, 1997), 154–200

J. Waldoff: 'Sentiment and Sensibility in *La vera costanza*', *Haydn Studies*, ed. W.D. Sutcliffe (Cambridge, 1998), 70–119

N. Miller: '*Commedia per musica*: Joseph Haydn Opera für Esterháza', in C. Dalhaus: *Europäische Romantik in der Musik*, i (Stuttgart, 1999), 243–353

L: Secular vocal

O.E. Deutsch: 'Haydns Kanons', *ZMw*, xv (1932–3), 112–24, 172

C. Hopkinson and C.B. Oldman: 'Thomson's Collections of National Song, with Special Reference to the Contributions of Haydn and Beethoven', *Edinburgh Bibliographical Society Transactions*, ii (1938–9), 1–64 [with thematic catalogue]; see also *Edinburgh Bibliographical Society Transactions*, iii (1949–51), 123–4; suppl. in 'Haydn's Settings of Scottish Songs in the Collections of Napier and Whyte', *Edinburgh Bibliographical Society Transactions*, iii (1949–51), 85–120

K. Geiringer: 'Haydn and the Folksong of the British Isles', *MQ*, xxxv (1949), 179–208

F. Grasberger *Die Hymnen Österreichs* (Tutzing, 1968)

A.P. Brown: 'Joseph Haydn and Leopold Hofmann's "Street Songs"', *JAMS*, xxxiii (1980), 356–83

O. Biba: *Gott erhalte! Joseph Haydns Kaiserhymne* (Vienna, 1982) [with facs. of first edn]

A. Riethmüller: 'Joseph Haydn und das Deutschlandlied', *AMw*, xliv (1987), 241–67

R. Bockholdt: '*Ein Mädchen, das auf Ehre hielt*: eine "sehr gewöhnliche Geschichte": der Erzählton in den Liedern und Instrumentalmusik von Joseph Haydn', *Liedstudien: Wolfgang Osthoff zum 60. Geburtstag*, ed. M. Just and R. Wiesend (Tutzing, 1989), 185–202

W. Kumbier: 'Haydn's English Canzonettas: Transformation in the Rhetoric of the Musical Sublime', *The Scope of Words: In Honor of Albert S. Cook*, ed. P. Barker, S.W. Goodwin and G. Handwerk (New York, 1991), 73–93

J. Rushton: 'Viennese Amateur or London Professional?: A Reconsideration of Haydn's Tragic Cantata *Arianna a Naxos*', *Music in Eighteenth-Century Austria: Cardiff 1991* (Cambridge, 1996), 232–45

A.P. Brown: 'Musical Settings of Anne Hunter's Poetry: from National Song to Canzonetta', *JAMS*, xlvii (1994), 39–89

B. Over: 'Arianna travestita: Haydns Kantate *Arianna a Naxos* in geistlichem Gewand', *Haydn-Studien*, vii/3–4 (1998), 384–97

M: Orchestral

H. Kretzschmar: 'Die Jugendsinfonien Joseph Haydns', *JbMP 1908*, 69–90

B. Rywosch: *Beiträge zur Entwicklung in Joseph Haydns Symphonik, 1759–1780* (Turbenthal, 1934)

A. Schering: 'Bemerkungen zu J. Haydns Programmsinfonien', *JbMP 1939*, 9–27; repr. in *Vom musikalischen Kunstwerk*, ed. F. Blume (Leipzig, 1949, 2/1951), 246–77

H.J. Therstappen: *Joseph Haydns sinfonisches Vermächtnis* (Wolfenbüttel, 1941)

H.C.R. Landon: *The Symphonies of Joseph Haydn* (London, 1955) [incl. thematic catalogues of authentic, spurious and doubtful syms., pp.605–823]; suppl. (London, 1961)

J.P. Larsen: 'The Symphonies', *The Mozart Companion*, ed. H.C.R. Landon and D. Mitchell (London and New York, 1956/R), 156–99

J.P. Larsen: 'Probleme der chronologischen Ordnung von Haydns Sinfonien', *Festschrift Otto Erich Deutsch*, ed. W. Gerstenberg, J. LaRue and W. Rehm (Kassel, 1963), 90–104

E.K. Wolf: 'The Recapitulations in Haydn's London Symphonies', *MQ*, lii (1966), 71–89

S. Gerlach: 'Die chronologische Ordnung von Haydns Sinfonien zwischen 1774 und 1782', *Haydn-Studien*, ii/1 (1969), 34–66

J.H. van der Meer: 'Die Verwendung der Blasinstrumente im Orchester bei Haydn und seinen Zeitgenossen', *Der junge Haydn: Graz 1970* (Graz, 1972), 202–20

L. Nowak: 'Die Skizzen zum Finale der Es-dur-Symphonie GA 99 von Joseph Haydn', *Haydn-Studien*, ii/3 (1970), 137–66

G. Thomas: 'Studien zu Haydns Tanzmusik', *Haydn-Studien*, iii/1 (1973), 5–28; see also *Jb für österreichische Kulturgeschichte*, ii: *Joseph Haydn und seine Zeit* (1972), 73–85

P.R. Bryan: 'The Horn in the Works of Mozart and Haydn', *Haydn Yearbook 1975*, 189–255

A. Hodgson *The Music of Joseph Haydn: the Symphonies* (London, 1976)

K. Marx: 'Über thematische Beziehungen in Haydns Londoner Symphonien', *Haydn-Studien*, iv/1 (1976), 1–19

M. Danckwardt: *Die langsame Einleitung: ihre Herkunft und ihr Bau bei Haydn und Mozart* (Tutzing, 1977)

B. Sponheuer: 'Haydns Arbeit am Finalproblem', *AMw*, xxxiv (1977), 199–224

J.P. Larsen: 'Zur Entstehung der österreichischen Symphonietradition (ca. 1750–1775)', *Haydn Yearbook 1978*, 72–80; Eng. trans. in Larsen, 1988

M.L. Martinez-Göllner: *Joseph Haydn: Symphonie Nr. 94* (Munich, 1979) [Meisterwerke der Musik, xvi]

R. Bard '"Tendenzen" zur zyklischen Gestaltung in Haydns Londoner Sinfonien', *GfMKB: Bayreuth 1981* (Kassel. 1984), 379–83

R. Bard: *Untersuchungen zur motivischen Arbeit in Haydns sinfonischem Spätwerk* (Kassel, 1982)

M.S. Cole: 'Haydn's Symphonic Rondo Finales: their Structural and Stylistic Evolution', *Haydn Yearbook 1982*, 113–42

N. Zaslaw: 'Mozart, Haydn and the *Sinfonia da Chiesa*', *JM*, i (1982), 95–124

K.-H. Schlager: *Joseph Haydn: Sinfonie Nr. 104 D-dur* (Munich, 1983) [Meisterwerke der Musik, xxxvii]

S. Gerlach: 'Haydns Orchesterpartituren: Fragen der Realisierung des Texts', *Haydn-Studien*, v/3 (1984), 169–83

R. Gwilt: 'Sonata-Allegro Revisited', *In Theory Only*, vii/5–6 (1984), 3–33

S.C. Fisher: *Haydn's Overtures and their Adaptations as Concert Orchestral Works* (diss., U. of Pennsylvania, 1985)

G. Schröder: 'Über das "klassische Orchester" und Haydns späte symphonische Instrumentation', *Musik-Konzepte*, no.41 (1985), 79–97

P. Benary: 'Die langsamen Einleitungen in Joseph Haydns Londoner Sinfonien', *Studien zur Instrumentalmusik: Lothar Hoffmann-Erbrecht zum 60. Geburtstag*, ed. A. Bingmann and others (Tutzing, 1988), 239–51

S. Gerlach: 'Haydns "chronologisische" Sinfonienliste für Breitkopf & Härtel', *Haydn-Studien*, vi/2 (1988), 116–29

H. Krones: '"Meine Sprache versteht man durch die ganze Welt: das "redende Prinzip" in Joseph Haydns Instrumentalmusik', *Wort und Ton im europäischen Raum: Gedenkschrift für Robert Schollum*, ed. H. Krones (Vienna, 1989), 79–108

S.A. Edgerton: *The Bass Part in Haydn's Early Symphonies: a Documentary and Analytical Study* (diss., Cornell U., 1990)

G. Feder: 'Joseph Haydns Konzerte: ihre Überlieferungs- und Wirkungsgeschichte', *Beiträge zur Geschichte des Konzerts: Festschrift Siegfried Kross*, ed. R. Emans and M. Wendt (Bonn, 1990), 115–24

D.P. Schroeder: *Haydn and the Enlightenment: the Late Symphonies and their Audience* (Oxford, 1990)

J.L. Schwartz: 'Periodicity and Passion in the First Movement of Haydn's "Farewell" Symphony', *Studies in Musical Sources and Style: Essays in Honor of Jan LaRue*, ed. E.K. Wolf and E.H. Roesner (Madison, WI, 1990), 293–338

E.R. Sisman: 'Haydn's Theater Symphonies', *JAMS*, xliii (1990), 292–352

J. Webster: 'The D-major Interlude in Haydn's "Farewell" Symphony', *Studies in Musical Sources and Style: Essays in Honor of Jan LaRue*, ed. E.K. Wolf and E.H. Roesner (Madison, WI, 1990), 339–80

J. Webster: 'On the Absence of Keyboard Continuo in Haydn's Symphonies', *EMc*, xviii (1990), 599–608

S.C. Fisher: 'Further Thoughts on Haydn's Symphonic Rondo Finales', *Haydn Yearbook 1992*, 85–107

P. Weber-Bockholdt: 'Joseph Haydns Sinfonien mit langsamen ersten Sätzen', *Mf*, xlv (1992), 152–61

A. Odenkirchen: *Die Konzerte Joseph Haydns: Untersuchungen zur Gattungstransformation in der zweiten Hälfte des 18. Jahrhunderts* (Frankfurt, 1993)

E. Haimo: *Haydn's Symphonic Forms: Essays in Compositional Logic* (Oxford, 1995)

A.P. Brown: 'The Sublime, the Beautiful and the Ornamental: English Aesthetic Currents and Haydn's London Symphonies', *Studies in Music History Presented to H.C. Robbins Landon*, ed. O. Biba and D.W. Jones (London, 1996), 44–71

S. Gerlach: 'Joseph Haydns Sinfonien bis 1774: Studien zur Chronologie', *Haydn-Studien*, vii/1–2 (1996), 1–287

M.E. Bonds: 'The Symphony as Pindaric Ode', *Haydn and his World*, ed. E. Sisman (Princeton, NJ, 1997), 131–53

W. Steinbeck: 'Die Konzertsatzform bei Haydn', *Traditionen – Neuansätze für Anna Amalie Abert (1906–1996)*, ed. K. Hortschansky (Tutzing, 1997), 493–518

R. Will: 'When God Met the Sinner, and Other Dramatic Confrontations in Eighteenth-Century Instrumental Music', *ML*, lxxviii (1997), 175–209

B. Harrison: *Haydn: the 'Paris' Symphonies* (Cambridge, 1998)

J. Webster: 'Haydn's Symphonies between *Sturm und Drang* and "Classical Style": Art and Entertainment', *Haydn Studies*, ed W.D. Sutcliffe (Cambridge, 1998), 218–45

N: Chamber without keyboard

A. Sandberger: 'Zur Geschichte des Haydn'schen Streichquartetts', *Altbayerische Monatsschrift*, ii (1900), 41–64; rev. in *Ausgewählte Aufsätze zur Musikgeschichte*, i (Munich, 1921/R), 224–65

D.F. Tovey: 'Franz Joseph Haydn', *Cobbett's Cyclopedic Survey of Chamber Music*, i (London, 1929–30, 2/1963/R), 514–48; repr. as 'Haydn's Chamber Music', *Essays and Lectures on Music* (London, 1943), 1–64

F. Blume: 'Joseph Haydns künstlerische Persönlichkeit in seinen Streichquartetten, *JbMP 1931*, 24–48; repr. in *Syntagma musicologicum: gesammelte Reden und Schriften* (Kassel, 1963), 526–51

W.O. Strunk: 'Haydn's Divertimenti for Baryton, Viola, and Bass', *MQ*, xviii (1932), 216–51

R. Sondheimer: *Haydn: a Historical and Psychological Study based on his Quartets* (London, 1951)

H.C.R. Landon: 'On Haydn's Quartets of Opera 1 and 2: Notes and Comments on Sondheimer's Historical and Psychological Study', *MR*, xiii (1952), 181–6

H.R. Edwall: 'Ferdinand IV and Haydn's Concertos for the "Lira Organizzata"', *MQ*, xlvii (1962), 190–203

W. Kirkendale: *Fuge und Fugato in der Kammermusik des Rokoko und der Klassik* (Tutzing, 1966; Eng. trans., enlarged, 1979)

L. Finscher: 'Joseph Haydn und das italienische Streichquartett', *AnMc*, no.4 (1967), 13–37

D. Bartha: 'Thematic Profile and Character in the Quartet Finales of Joseph Haydn (a Contribution to the Micro-Analysis of Thematic Structure)', *SMH*, xi (1969), 35–62

I. Saslav: *Tempos in the String Quartets of Joseph Haydn* (diss., Indiana U., 1969)

H. Unverricht: *Geschichte des Streichtrios* (Tutzing, 1969)

H. Unverricht: 'Zur Chronologie der Barytontrios von Joseph Haydn', *Symbolae historiae musicae: Hellmut Federhofer zum 60. Geburtstag*, ed. F.W. Riedel and H. Unverricht (Mainz, 1971), 180–89

L. Somfai: '"Ich war nie ein Geschwindschreiber ... ": Joseph Haydns Skizzen zum langsamen Satz des Streichquartetts Hoboken III:33', *Festskrift Jens Peter Larsen*, ed. N. Schiørring, H. Glahn and C.E. Hatting (Copenhagen, 1972), 275–84

L. Finscher: *Studien zur Geschichte des Streichquartetts*, i: *Die Entstehung des klassischen Streichquartetts: von den Vorformen zur Grundlegung durch Joseph Haydn* (Kassel, 1974)

O. Moe: 'Texture in Haydn's Early Quartets', *MR*, xxxv (1974), 4–22

J. Webster: 'Towards a History of Viennese Chamber Music in the Early Classical Period', *JAMS*, xxvii (1974), 212–47

J. Webster: 'The Chronology of Haydn's String Quartets', *MQ*, lxi (1975), 17–46

J. Webster: 'Freedom of Form in Haydn's Early String Quartets', *Haydn Studies: Washington DC 1975* (New York, 1981), 522–30

F. Salzer: 'Haydn's Fantasia from the String Quartet, Opus 76, No.6', *Music Forum*, iv (1976), 161–94

J. Webster: 'Violoncello and Double Bass in the Chamber Music of Haydn and his Viennese Contemporaries, 1750–1780', *JAMS*, xxix (1976), 413–38

J. Webster: 'The Bass Part in Haydn's Early String Quartets', *MQ*, lxiii (1977), 390–424

T. Crawford: 'Haydn's Music for Lute', *Le luth et sa musique II: Tours 1980* (Paris, 1980) 69–86

L. Somfai: 'An Introduction to the Study of Haydn's String Quartet Autographs', *The String Quartets of Haydn, Mozart and Beethoven. Cambridge, MA, 1979* (Cambridge MA, 1980), 5–51

J. Webster: 'The Significance of Haydn's String Quartet Autographs for Performance Practice', ibid., 62–95

L. Somfai: 'Opus-Planung und Neuerung bei Haydn', *SMH*, xxii (1980), 87–110

W. Konold: 'Normerfüllung und Normverweigerung beim späten Haydn am Beispiel des Streichquartetts op.76 Nr.6', *Joseph Haydn: Cologne 1982* (Regensburg, 1985), 54–73

E. Kubitschek: 'Die Flötentrios Hob. IV:6–11', *Joseph Haydn: Vienna 1982* (Munich, 1986), 419–26

E.R. Sisman: 'Haydn's Baryton Pieces and his Serious Genres', ibid., 426–35

B.S. Brook: 'Haydn's String Trios: a Misunderstood Genre', *CMc*, no.36 (1983), 61–77

C.A. Gartrell: *The Baryton: the Instrument and its Music* (diss., U. of Surrey, 1983) [chap. 13, Esterházy]

W. Steinbeck: 'Mozart's "Scherzi": zur Beziehung zwischen Haydns Streichquartetten op.33 und Mozarts Haydn-Quartetten', *AMw*, xli (1984), 208–31

H. Keller: *The Great Haydn Quartets: their Interpretation* (London, 1986)

J. Neubacher: *Finis Coronat Opus: Untersuchungen zur Technik der Schlussgestaltung in der Instrumentalmusik Joseph Haydns, dargestellt am Beispiel der Streichquartette* (Tutzing, 1986)

L. Somfai: '"Learned Style" in Two Late String Quartet Movements of Haydn', *SMH*, xxviii (1986), 325–49

S.E. Tepping: *Fugue Process and Tonal Structure in the String Quartets of Haydn, Mozart, and Beethoven* (diss., Indiana U., 1987)

M. Bandur: *Form und Gehalt in den Streichquartetten Joseph Haydns: Studien zur Theorie der Sonatenform* (Pfaffenweiler, 1988)

J. Webster: 'Haydns frühe Ensemble-Divertimenti: Geschlossene Gattung, meisterhafter Satz', *Gesellschaftsgebundene instrumentale Unterhaltungsmusik des 18. Jahrhunderts: Eichstätt 1988*, 87–103

N. Schwindt-Gross: *Drama und Diskurs: zur Beziehung zwischen Satztechnik und motivischem Prozess am Beispiel der durchbrochenen Arbeit in den Streichquartetten Haydns und Mozarts* (Laaber, 1989)

G. Edwards: 'The Nonsense of an Ending: Closure in Haydn's String Quartets', *MQ*, lxxv (1991), 227–54

A.P. Brown: 'Haydn and Mozart's 1773 Stay in Vienna: Weeding a Musicological Garden', *JM*, x (1992), 192–230

W.D. Sutcliffe: *Haydn: String Quartets, Op.50* (Cambridge, 1992)

M.E. Bonds: 'The Sincerest Form of Flattery? Mozart's "Haydn" Quartets and the Question of Influence', *Studi musicali*, xxii (1993), 365–409

G.J. Winkler: 'Opus 33/2: zur Anatomie eines Schlusseffekts', *Haydn-Studien*, vi/4 (1994), 288–97

H. Danuser: 'Das Ende als Anfang: Ausblick von einer Schlussfigur bei Joseph Haydn', *Studien zur Musikgeschichte: eine Festschrift für Ludwig Finscher*, ed. A. Laubenthal and K. Kusan-Windweh (Kassel, 1995), 818–27

H. Walter: 'Zum Wiener Streichquartett der Jahre 1780 bis 1800', *Haydn-Studien*, vii/3–4 (1998), 289–314

W. Drabkin: *A Reader's Guide to Haydn's Early Quartets* (Westport CT and London, 2000)

O: Keyboard

NewmanSCE

H. Abert: 'Joseph Haydns Klavierwerke', *ZMw*, ii (1919–20), 553–73; 'Joseph Haydns Klaviersonaten', iii (1920–21), 535–52

H. Schenker: 'Haydn: Sonate Es Dur' [H XVI:52], *Der Tonwille*, i (1922), 3–21

H. Schenker: 'Haydn: Sonate C Dur' [H XVI:50], *Der Tonwille*, ii (1923), 15–18

H. Schenker: 'Das Organische der Sonatenform', *Meisterwerk in der Musik*, ii (1926/R), 43–54; Eng. trans., *JMT*, xii (1968); Eng. trans., as *Masterwork in Music* [H XVI:44, first movement]

E.F. Schmid: 'Joseph Haydn und die Flötenuhr', *ZMw*, xiv (1931–2), 193–221

H. Schenker: *Fünf Urlinie-Tafeln* (New York, 1933; Eng. trans., 1969, with new introduction and glossary by F. Salzer as *Five Graphic Music Analyses*) [Piano Sonata H XVI:49]

W.O. Strunk: 'Notes on a Haydn Autograph', *MQ*, xx (1934), 192–205

D.F. Tovey: 'Haydn: Pianoforte Sonata in E flat, No.1', *Essays in Musical Analysis: Chamber Music* (London, 1944/R), 93–105 [H XVI:52]

G. Feder: 'Probleme einer Neuordnung der Klaviersonaten Haydns', *Festschrift Friedrich Blume*, ed. A.A. Abert and W. Pfannkuch (Kassel, 1963), 92–103

F. Eibner: 'Die authentische Klavierfassung von Haydns Variationen über "Gott erhalte"', *Haydn Yearbook 1970*, 281–306

G. Feder: 'Haydns frühe Klaviertrios: eine Untersuchung zur Echtheit und Chronologie', *Haydn-Studien*, ii/4 (1970), 289–316

G. Feder: 'Wieviel Orgelkonzerte hat Haydn geschrieben?', *Mf*, xxiii (1970), 440–44

E.F. Schmid: 'Neue Funde zu Haydns Flötenuhrstücken', *Haydn-Studien*, ii/4 (1970), 249–55

H. Walter: 'Haydns Klaviere', *Haydn-Studien*, ii/4 (1970), 256–88

H. Walter: 'Das Tasteninstrument beim jungen Haydn', *Der junge Haydn: Graz 1970* (Graz, 1972), 237–48

A.P. Brown: 'The Structure of the Exposition in Haydn's Keyboard Sonatas', *MR*, xxxvi (1975), 102–29

B. Wackernagel: *Joseph Haydns frühe Klaviersonaten: ihre Beziehungen zur Klaviermusik um die Mitte des 18. Jahrhunderts* (Tutzing, 1975)

L. Steinberg: *Sonata Form in the Keyboard Trios of Joseph Haydn* (diss., New York U., 1976)

L. Somfai: *Joseph Haydn zongoraszonátái hangszerválasztás és előadói gyakorlat, mufaji tipológia és stílselemzés* (Budapest, 1979; Eng. trans., with C, Greenspan, 1995, as *The Keyboard Sonatas of Joseph Haydn: Instruments and Performance Pactice, Genres and Styles*)

P. Badura-Skoda: 'Beiträge zu Haydns Ornamentik', *Musica*, xxxvi (1982), 409–18; Eng. trans., *Piano Quarterly*, no.134 (1986), 38–48

M. Fillion: *The Accompanied Keyboard Divertimenti of Haydn and his Viennese Contemporaries (c. 1750–1780)* (diss., Cornell U., 1982)

W.S. Newman: 'Haydn as Ingenious Exploiter of the Keyboard', *Joseph Haydn: Vienna 1982* (Regensburg, 1985), 43–53

A.W.J.G. Ord-Hume: *Joseph Haydn and the Mechanical Organ* (Cardiff, 1982)

R. Kamien: 'Aspects of Motivic Elaboration in the Opening Movement of Haydn's Piano Sonata in C♯ minor', *Aspects of Schenkerian Theory*, ed. D. Beach (New Haven, CT, 1983), 77–93

J. Neubacher: '"Idee" und "Ausführung": zum Kompositionsprozess bei Joseph Haydn', *AMw*, xli (1984), 187–207

B. Shamgar: 'Rhythmic Interplay in the Retransitions of Haydn's Piano Sonatas', *JM*, iii (1984), 55–68

A.P. Brown: *Joseph Haydn's Keyboard Music: Sources and Style* (Bloomington, IN, 1986)

K. Komlós: 'Haydn's Keyboard Trios Hob. XV:5–17: Interaction between Texture and Form', *SMH*, xxviii (1986), 351–400

K. Komlós: 'The Viennese Keyboard Trio in the 1780s: Sociological Background and Contemporary Reception', *ML*, lxviii (1987), 222–34

S.P. Rosenblum: *Performance Practices in Classic Piano Music* (Bloomington, IN, 1988)

F. Krummacher: 'Klaviertrio und sinfonischer Satz: zum Adagio aus Haydns Sinfonie Nr.102', *Quaestiones in musica: Festschrift für Franz Krautwurst*, ed. F. Brusniak and H. Leuchtmann (Tutzing, 1989), 325–35

W.D. Sutcliffe: *The Piano Trios of Haydn* (diss., U. of Cambridge, 1989)

U. Leisinger: *Joseph Haydn und die Entwicklung des klassischen Klavierstils bis ca. 1785* (Laaber, 1994)

W. Petty: 'Cyclic Integration in Haydn's E♭ Piano Sonata Hob. XVI:38', *Theory and Practice*, xix (1994), 31–55

J. Brauner: *Studien zu den Klaviertrios von Joseph Haydn* (Tutzing, 1995)

K. Komlós: *Fortepianos and their Music: Germany, Austria and England* (Oxford, 1995)

T. Beghin: 'Haydn as Orator: a Rhetorical Analysis of his Keyboard Sonata in D major, Hob. XVI:42', *Haydn and his World*, ed. E. Sisman (Princeton, NJ, 1997), 201–54

B. Harrison: *Haydn's Keyboard Music: Studies in Performance Practice* (Oxford, 1997)

J. Webster: 'The Triumph of Variability: Haydn's Articulation Markings in the Autograph of Sonata No.49 in E-flat', *Haydn, Mozart, & Beethoven: Essays in Honour of Alan Tyson*, ed. S. Brandenburg (Oxford, 1998), 33–64

P: Reputation

L. de La Laurencie: 'L'apparition des oeuvres d'Haydn à Paris', *RdM*, xiii (1932), 191–205

M.D.H. Norton: 'Haydn in America (before 1820)', *MQ*, xviii (1932), 309–37

L. Schrade: 'Das Haydn-Bild in den ältesten Biographien', *Die Musikerziehung*, ix (1932), 163–9, 200–13, 244–9

A. Sandberger: 'Zur Einbürgerung der Kunst Josef Haydns in Deutschland', *NBeJb 1935*, 5–25

F. Lesure: 'Haydn en France', *Konferenz zum Andenken Joseph Haydns: Budapest 1959* (Budapest, 1961), 79–84

K.G. Fellerer: 'Zum Joseph-Haydn-Bild im frühen 19. Jahrhundert', *Anthony van Hoboken: Festschrift*, ed. J. Schmidt-Görg (Mainz, 1962), 73–86

A. Palm: 'Unbekannte Haydn-Analysen', *Haydn Yearbook 1968*, 169–94 [analyses by Momigny]

M.S. Cole: 'Momigny's Analysis of Haydn's Symphony no.103', *MR*, xxx (1969), 261–84

B. Steinpress: 'Haydns Oratorien in Russland zu Lebzeiten des Komponisten', *Haydn-Studien*, ii/2 (1969), 77–112

C. Höslinger: 'Der überwundene Standpunkt: Joseph Haydn in der Wiener Musikkritik des 19. Jahrhunderts', *Jb für österreichische Kulturgeschichte*, i/2: *Beiträge zur Musikgeschichte des 18. Jahrhunderts* (1971), 116–42

A.P. Brown: 'The Earliest English Biography of Haydn', *MQ*, lix (1973), 339–54

I. Lowens: 'Haydn in America', *Haydn Studies: Washington DC 1975* (New York, 1981), 35–48

R. Stevenson: 'Haydn's Iberian World Connections', *Inter-American Music Review*, iv/2 (1981–2), 3–30

G. Feder: 'Joseph Haydn 1982: Gedanken über Tradition und historische Kritik', *Joseph Haydn: Vienna 1982* (Munich, 1986), 597–611

U. Tank: 'Joseph Haydns geistliche Musik in der Anschauung des 19. Jahrhunderts', *Joseph Haydn: Cologne 1982* (Regensburg, 1985), 215–62

D. Wyn Jones: 'Haydn's Music in London in the Period 1760–1790', *Haydn Yearbook 1983*, 144–72

M.S. Morrow: *Concert Life in Haydn's Vienna: Aspects of a Developing Musical and Social Institution* (New York, 1989); see also review by D. Edge, *Haydn Yearbook 1992*, 108–66

G.A. Wheelock: 'Marriage à la Mode: Haydn's Instrumental Works "Englished" for Voice and Piano', *JM*, viii (1990), 357–97

T. Tolley: 'Music in the Circle of Sir William Jones: a Contribution to the History of Haydn's Early Reception', *ML*, lxxiii (1992), 525–50

S. McVeigh: *Concert Life in London from Mozart to Haydn* (Cambridge, 1993)

H. Irving: 'William Crotch on "The Creation"', *ML*, lxxv (1994), 548–60

L. Botstein: 'The Demise of Philosophical Listening: Haydn in the 19th Century', *Haydn and his World*, ed. E. Sisman (Princeton, NJ, 1997), 255–85

L. Botstein: 'The Consequences of Presumed Innocence: the Nineteenth-century Reception of Joseph Haydn', *Haydn Studies*, ed. W.D. Sutcliffe (Cambridge, 1998), 1–34

INDEX

1) 1 violin line
Rising theme.
pulsating cello part
more only except
of b 4-bar forward

2) major key swaying
tod — more agitated
minor key chromatic
shadowy cellos
cellos into by shadows

3) lush sweet luscious.
1st v again — climactic.

4) furmost Tough. Fierce
sombre gruff